LED
Lighting
Explained

Understanding LED Sources,
Fixtures, Applications,
and Opportunities

Text by **Jonathan Weinert,** Philips Color Kinetics
Illustrations by **Charles Spaulding,** Philips Color Kinetics

Contents

1	Introducing LED Lighting 3
	LED Lighting Fixtures: A New Kind of Light 6
	LED Lighting Installations: From Simple to Complex 7
	LEDs and Everyday Illumination 8
	Retrofitting Boston's Marriott Custom House Tower: A Case Study 8
	LEDs and the Green Revolution 10
	The Advantages of LED Lighting 11

2	LED Basics 15
	A Brief History of LEDs 15
	How LEDs Work 16
	LED Anatomy 17
	How LEDs Produce Different Colors 18
	How LEDs Produce Millions of Colors 19
	How LEDs Produce White Light 20
	LED Fixture Anatomy 21

3	LED Lighting in Detail 23
	Evaluating Light Output: The Importance of Delivered Light 24
	The Trouble with Lumens 26
	What Exactly Is a Lumen? 27
	Deficiencies of the Eye-Sensitivity Curve 29
	Relative Photometry, Absolute Photometry, and Efficiency . . . 31
	Lensing, Filtering, Shading, and Other Sources of Loss 32
	General Lighting Example: The Downlight 33
	Cove Lighting Example: Directional Light 34
	Quality of Light 35
	Color-Rendering Index and White-Light LEDs 36
	Do LED Light Sources Produce Acceptable CRI? 36
	Shortcomings of CRI for White-Light LEDs 36
	LEDs and Color Consistency 37

Understanding Correlated Color Temperature. 38
Consistent, Reliable Color: All About Binning 39
Choosing the Right White 42
Delivering the Whole Range of Color Temperatures. 43
Efficacy of LED Lighting Fixtures 44
Comparing the Efficiency of LED Lighting Fixtures
and Conventional Lighting Fixtures 44
Some Real-World Examples. 44
Minimizing Off-State Power Consumption 45
The Importance of Thermal Management 45
About Junction Temperature. 46
How Junction Temperature Affects Light Output 47
How Junction Temperature Affects Useful Life 48
Useful Life: Understanding LM-80, Lumen Maintenance,
and LED Fixture Lifetime. 48
Rated Lamp Life of Conventional Sources. 49
Lumen Maintenance and Lumen Depreciation 49
Defining the Useful Life of LED Light Sources. 51
The Lumen Maintenance Gap 51
The Useful Life of LED Sources in Lighting Fixtures 54
Useful Life Is Not Fixture Lifetime 54
Comparing the Useful Life of Conventional Lamps
and LED Lighting Fixtures. 55
Getting Dependable, Accurate Information 56
Driving and Powering LED Lighting Fixtures 57
LED Drivers. 58
Power Options for LED Lighting Fixtures 58
Low-Voltage Power Distribution. 58
Onboard Power Integration 59
Inboard Power Integration 60
Controlling LED Lighting Fixtures 61
DMX Control 61
Ethernet Control 62
Other Control Options. .64
Dimming LED Lighting Fixtures. 64
Dimming LED Lighting Fixtures Via DMX or Another
Control Interface 64
Dimming LED Lighting Fixtures with Commercially
Available Dimmers. 65
Dimming Threshold and Dimmer Wattage 66

4   LED Lighting Applications. 69
    Task Lighting 69
        eW Profile Powercore 70
        Case Study: Under-Cabinet Lighting / Private Residence 70
    Downlighting 71
        eW Downlight Powercore 72
        Calculite LED Downlight 72
        Case Study: Downlighting / Retail Space. 73
    Cove Lighting. 74
        eW Cove QLX Powercore 74
        iW Cove Powercore 75
        iColor Cove MX Powercore .75
        Case Study: Cove Lighting / Historical Landmark 76
        Case Study: Cove Lighting / Hospitality Space. 76
    Wall Washing 78
        ColorBlast Powercore. 78
        ColorBlaze 78
        Case Study: Wall Washing / Contemporary Landmark. 79
    Wall Grazing 79
        ColorGraze Powercore 79
        Case Study: Wall Grazing / Private Residence 80
    Floodlighting 81
        ColorReach Powercore 81
        Case Study: Exterior Architectural Floodlighting / Public
        Building 81
    Roadway and Area Lighting 83
        Radiant 83
    Safety and Utility Lighting. 84
        Philips Gardco Crosswalk System 84
    Accent Lighting 86
        ColorBurst 6 86
        iColor MR g2 86
        C-Splash 2. 87
        Case Study: Accent Lighting / Hospitality Space 87
        Case Study: Accent Lighting / Public Interior 88
    Direct View Lighting 89
        iColor Accent Powercore 90
        iColor Flex LMX 91
        iColor Tile MX 91
        Case Study: Large-Scale Video Display / Exterior
        Architecture 91

5   Doing Business with LED Lighting .93
         Global LED Lighting Fixture Market Size 94
         Driving the Demand: Legislation, Policies, and Incentives 95
             EU-27 Environmental Lighting Initiatives 96
                 Waste Electrical and Electronic Equipment
                 Directive (WEEE) 96
                 Restriction of Hazardous Substances
                 Directive (RoHS). 97
                 Ecodesign Directive for Energy-Related
                 Products (ErP) 97
                 Energy Performance of Buildings Directive (EPBD) 99
             North American Environmental Lighting Initiatives 99
                 The Energy Policy Act (EPAct) of 2005 100
                 The Energy Independence and Security Act (EISA)
                 of 2007 100
                 California's Title 24 100
                 ENERGY STAR for Solid State Lighting Luminaires 101
                 Leadership in Energy and Environmental
                 Design (LEED) 101
             Environmental Lighting Initiatives Around the World 102
         Making the Business Case 103
             Total Cost of Ownership Example. 104
             Basic Lighting Economics Example 105

Notes 107

Glossary. 113

1
Introducing LED Lighting

You MAY NOT know it, but LED lighting is already in use in an astonishing array of locations and applications. Color-changing LED lighting is currently illuminating architectural landmarks and signature buildings worldwide, from the CN Tower in Toronto, Canada to the Bosphorus Bridge over the Istanbul Strait in Turkey. Full-color LED lighting is in extensive use in theatres, concerts, stage productions, restaurants, casinos, and a variety of public spaces calling for dramatic and dynamic lighting displays.

Fixed-color and white-light LED fixtures support the full range of outdoor lighting applications, including road and tunnel lighting, sports and arena lighting, streetlighting and area lighting, landscape lighting, signage, and airport runway lighting. Well-designed retrofit lamps in a variety of form factors deliver

Color-changing LED lighting systems are in use around the world. LED lighting dramatically and dynamically illuminates the Avenue of the Arts in Philadelphia, Pennsylvania.

many of the benefits of LED technology to existing lighting systems and infrastructures. LED white light is beginning to make extensive inroads in general illumination. White-light LED fixtures in widespread use include cove lights, task lights, and downlights for retail spaces, museums, offices, schools, laboratories, hospitals, and private residences.

Well-designed LED lighting fixtures offer features, performance, and economy equivalent or superior to comparable conventional lighting fixtures. LED lighting fixtures can:

- Direct abundant useful light to tasks, work surfaces, and target areas in most lighting applications, both indoors and outdoors

- Reliably deliver consistent, high-quality colored and white light with virtually no visible color variation from fixture to fixture

White-light LEDs have come into their own for general illumination. LED cove lighting gives an elegant and inviting look to historic Old North Church in Boston, Massachusetts.

- Support almost any lighting application

- Provide a full spectrum of solid colors, and white light in the full range of shades from warm to cool

- Offer full-color and tunable white light for dynamic color-changing effects, intricate light shows, and large-scale video displays impossible to accomplish with conventional lighting fixtures

- Meet or exceed the energy efficiency of comparable conventional lighting fixtures

- Reliably deliver useful light for many thousands of hours after fluorescent and incandescent sources have failed
- Offer ease of installation and operation in new and retrofit applications, using familiar techniques and tools
- Resist damage caused by vibration, and operate efficiently in cold temperatures
- Lower the total cost of light through energy efficiency, durability, minimal maintenance requirements, and low failure rates, often achieving payback in less than one year

LED lighting is leading the lighting industry in its ongoing effort to develop truly green, sustainable solutions and market them affordably. As costs come down, new standards, green initiatives, and legislation are creating an enormous opportunity for the adoption of LED lighting both nationally and internationally.

One remaining barrier to the widespread adoption of LED lighting is lack of familiarity with the technology. This handbook is intended to remove that barrier by providing an overview of current LED lighting technology, applications, and business opportunities.

LED Fixtures for the Full Range of Lighting Applications

LED-based solutions are available today to support virtually any lighting application, including these ten major application areas. See Chapter 4 for more details on applications, featured LED lighting fixtures, and case studies of successful LED-based installations around the world.

Task Lighting   Downlighting   Cove Lighting

Wall Washing   Wall Grazing   Direct View Lighting

Floodlighting   Roadway Lighting   Safety / Utility Lighting   Accent Lighting

LED Lighting Fixtures: A New Kind of Light

LED lighting fixtures are as easy to install, operate, and maintain as conventional lighting fixtures, but they also differ from conventional fixtures in some significant ways. LED lighting fixtures are:

- Solid-state devices that use semiconductor-based light sources and other electronics — unlike incandescent fixtures, which use glass lamps and filaments, or fluorescent sources, which use electric energy to excite gases inside the lamps

- Inherently digital devices, many of which can be precisely controlled and dimmed using digital controllers

- Integrated systems that erase the distinction between lamp and luminaire. In many LED fixtures, the "lamps" — the LEDs — are inseparable from the fixture.

Photo courtesy of Electroland

Dynamic LED lighting fixtures bring the drama of theatrical lighting to public and residential spaces. Color-changing and tunable white light LED fixtures exceed the capabilities of conventional lighting fixtures, opening up new vistas in city beautification and architectural, accent, and decorative lighting. Here, iColor® Cove MX Powercore LED fixtures from Philips Color Kinetics create an immersive, interactive environment for Target at Rockefeller Center in New York.

- Often part of a system that includes power / data supplies, controllers, and cabling

- Inherently directional, making them more efficient than fixtures which project light in all directions. Lensing, positioning, and aiming features are integrated into most LED lighting fixtures.

- The only light sources that *increase* in efficacy when dimmed

Fixtures can contain clusters of white or colored LEDs. Fixtures with white-light LEDs can produce one shade of cool, neutral, or warm white light, while tunable white fixtures combine cool and warm white LEDs to produce a range of shades that can be controlled with simple push-button devices. Fixtures with solid-color LEDs produce light of a fixed hue — green, blue, royal blue, amber, and so on. Color-changing (RGB) LED fixtures combine red, green, and blue LEDs to produce millions of colors and dynamic color-changing effects. Multi-spectrum LED fixtures use additional colors (RGB plus channels of amber and white, for example) to alter or increase the range of possible colors.

LED fixtures can be powered by an external power supply or through direct line voltage. Some direct line voltage fixtures integrate power management directly into the fixture's electronics, eliminating the need for bulky and inefficient external power supplies. Some line-voltage fixtures increase fixture efficacy by reducing or eliminating off-state power consumption.

LED Lighting Installations: From Simple to Complex

While large-scale LED lighting installations can be very intricate and complex, many white light or solid color installations for everyday or accent illumination are as simple as similar installations using incandescent or fluorescent sources.

The simplest installations — white-light LED downlights or under-cabinet lights, for example — plug into standard outlets or connect directly to mains power, just like conventional lighting fixtures. Linear lighting fixtures, such as under-cabinet and cove lights, typically feature integrated end-to-end connectors and pre-configured leader and jumper cables.

Complex systems consisting of multiple runs of intelligent white or color-changing LED fixtures include power supplies and controllers to control, synchronize, and network the fixtures for a range of lighting effects and dynamic, color-changing light shows.

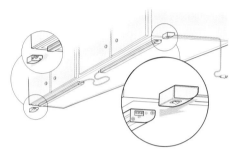

Installing LED fixtures can be as simple and straightforward as installing conventional lighting fixtures. For example, eW® Profile Powercore from Philips Color Kinetics is a simple white-light under-cabinet fixture that can plug directly into a standard outlet. Integrated end-to-end connectors and pre-configured cables further simplify installation.

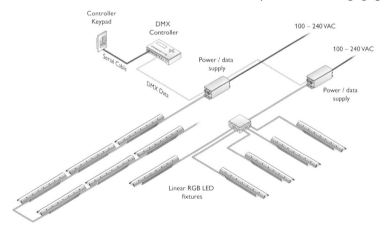

Complex LED-based installations can feature multiple series of RGB LED fixtures, a digital controller, and one or more power / data supplies The linear wall-washing fixtures shown here can be individually controlled to display intricate, color-changing light shows and effects.

LEDs and Everyday Illumination

Recent advances in LED technology have made white-light LED lighting fixtures viable, attractive, and, in an increasing number of situations, preferable for everyday illumination in public and private spaces.

The energy efficiency of white-light LED fixtures now approaches and in some cases surpasses that of conventional incandescent and CFL sources. The light output and quality of the best white-light LED sources also compare favorably with many conventional sources. Advances in energy efficiency, light output, and quality of light promise to make LEDs increasingly desirable over time.

Well-designed LED lighting fixtures offer superior light mixing and uniformity, especially in cove, under-cabinet, and accent lighting applications — applications where current LED lighting solutions are especially strong. The LEDs in properly operated lighting fixtures typically maintain effective light output for 50,000 hours or more, which makes them advantageous in situations where lamp and fixture maintenance is problematic.

LED lighting installations are cost-effective. While initial costs can exceed initial costs for comparable conventional installations, dramatically reduced labor, maintenance, energy, and replacement costs lower the total cost of light. LED lighting systems consistently achieve payback in less than three years — in less than one year, in some cases — and both the initial and total costs of LED lighting systems are expected to drop over time.

A growing number of laws, incentives, and standards aimed at increasing the energy efficiency and decreasing the environmental impact of lighting systems favor the use of LED lighting systems in both new and retrofit applications. Some LED lighting fixtures currently achieve levels of energy efficiency consistent with compliance ratings such as California's Title 24 and ENERGY STAR in North America and the Ecodesign Directive for Energy-Related Products in the EU. Emerging standards and new testing procedures are yielding legitimate, non-biased measurements of LED lighting fixture capabilities, and are paving the way for more accurate product comparisons between LED and traditional lighting sources.

Retrofitting Boston's Marriott Custom House Tower: A Case Study

Boston's first skyscraper, the Marriott Custom House Tower, was completed in 1915. In 2008, the tower underwent a long-awaited lighting redesign that restores its prominence on the Boston skyline in a smart and sustainable way. Formerly lit by PAR 38 halogen spotlights, the tower was restored using energy-efficient, low-maintenance white light LED fixtures from Philips Color Kinetics. A combination of 125 eW® Graze Powercore and eW Blast Powercore fixtures now fully illuminate the tower from the 17th floor to the peak.

The LED fixtures in the installation consume just one third the energy of the previous incandescent sources. At six hours of use per day, they are expected

eW Blast Powercore   eW Graze Powercore

to maintain useful light output for more than 20 years, dramatically reducing replacement and maintenance costs.

Powercore®, a proprietary power management technology, allows the fixtures to directly accept line voltage, eliminating the need for special cabling and external low-voltage power supplies. Because the LED fixtures use conventional wiring and power sources, the lighting designers were able to easily replace the former incandescent light fixtures one-for-one in their existing locations and mountings.

The long useful source life and efficiency of white-light LED lighting fixtures open up new possibilities for sustainable architectural lighting. As the Custom House Tower installation demonstrates, LED lighting has arrived as a viable, energy-efficient alternative for general illumination.

LEDs and the Green Revolution

According to the Next Generation Lighting Industry Alliance (NGLIA), lighting represents over 20% of all electricity use in the U.S.[1] At the average price of electricity in 2008, lighting accounts for about $60 billion per year in energy costs.[2,3] Experts predict that the use of LEDs could reduce lighting energy by 50% by the year 2025, which could have a potentially transforming effect on the U.S. economy.[4] Simply replacing all domestic incandescent traffic lights with LEDs would save hundreds of millions of dollars a year.[5] Widespread LED lighting use could potentially save 189 terawatt hours (TWh) per year, eliminating the annual output of about 30 1,000-megawatt power plants.[6]

Epic-scale color-changing effects, courtesy of ColorBlast® 12 LED spotlights from Philips Color Kinetics, wash the 120-foot-high facade of the Hard Rock Hotel in Las Vegas, Nevada. The LED-based system is extremely energy efficient, drawing over $16,000 less per year in electricity than the previous metal halide system.

Recent legislation and energy savings initiatives, such as the EPA's ENERGY STAR, California's Title 24, and the U.S. Green Building Council's LEED program, favor the use of LEDs over conventional light sources. In the US, the Energy Independence and Security Act of 2007 looks to reduce residential lighting energy by 50% and commercial lighting energy by 25% before 2018. Ban the Bulb legislation may ban the sale of most current incandescent bulbs worldwide by 2014. Incentives such as the Energy Policy Act of 2005, which offers tax incentives for energy-saving technologies, and the DoE-sponsored L-Prize, which offers a $10 million prize to the first manufacturer to produce a viable low-energy replacement for the 60-watt incandescent light bulb, drive the continuing research, development, and manufacturing of LED-based alternatives.

The long useful life of LED sources means fewer fixture replacements and therefore fewer lamps to discard. LED sources and fixtures contain no mercury, so discards do not need to be treated as hazardous waste, as do fluorescent lamps and tubes. Leading LED fixture manufacturers, such as Philips Color Kinetics, are using close to 100% recyclable materials and components in their fixtures, and are engineering their fixtures so that they can be disassembled for recycling purposes.

LED lighting is poised to expand even more aggressively from its base in accent and effect lighting into general illumination applications. In the US, the Obama administration has offered unequivocal support for energy-efficient lighting alternatives in general, and for LED lighting in particular. In a January 8, 2009, economic recovery and reinvestment speech, President Obama called for improving the energy-efficiency of more than 75% of federal buildings and two million American homes.[7] After a July 2, 2009 meeting with Chuck Swoboda, the CEO of Cree, a leading U.S. manufacturers of LEDs, Mr. Obama pointed to the potential of LED lighting to "save a huge amount on energy costs."[8]

The Advantages of LED Lighting

Despite the fact that LED lighting is a relatively new technology, LED lighting sources already equal and in many cases surpass the efficiency, quality, cost-effectiveness, and environmental friendliness of conventional lighting sources. LED lighting sources outstrip incandescent sources for virtually all applications, and HID sources for colored light applications.

LED lighting can't do everything — yet. LED sources are making headway in some white light applications as replacements for HID and linear fluorescent sources, but they still have a way to go before their widespread adoption for general illumination applications. Nevertheless, LED lighting systems today afford a number of advantages over conventional systems:

- LED lighting can be **five times more energy-efficient** than incandescent and halogen sources, and comparable in efficacy to CFLs. LED sources continue to improve and are quickly gaining on the energy efficiency of fluorescent tubes.

- LED lighting **fixtures are directional, throwing light only where it's needed**. LEDs have a much smaller light source than CFLs, allowing more efficient optics and better control of light.

- The **quality of white-light LEDs is now comparable to CFLs, HIDs, and linear fluorescents**. Recent advances in LED manufacturing techniques assure consistency in color and color temperature comparable to or better than traditional sources.

- LED sources offer a **significantly longer useful life than conventional light sources**. This means less replacement and maintenance. For instance, halogen lamps may have to be replaced from 12 to 20 times before a comparable, properly designed LED alternative must be replaced once.

- Unlike conventional sources, **LED sources continue to be useful even after their light output has decreased significantly**. Outright failure of LED sources is rare.

- The amount of light output from LED sources has been improving by 35% each year since the LED was invented. At the same time, the cost of LEDs has been dropping by 20% each year for several decades. This means **the overall performance of LEDs is doubling about every 18 to 24 months**.[9]

- Because LEDs **do not emit infrared radiation**, they can be installed in heat-sensitive areas, near people and materials, and in tight spaces where conventional sources could be dangerous.

- Unlike fluorescent sources, LEDs **do not emit harmful UV rays** that can degrade materials or fade paints and dyes, making them ideal for use in retail displays, museums, and art galleries.

- LED fixtures do generate heat, but **the beam of light from an LED fixture is cool**. LED fixtures with well-designed thermal management features shield users from excessive or harmful heat.

- LED light sources can operate at colder temperatures and withstand impact and vibrations, making them **suitable for extreme environments** or

The multi-layered atrium of the World Market Center in Las Vegas features an extensive labyrinth of coves. Thousands of linear feet of eW Cove Powercore, an LED cove fixture from Philips Color Kinetics, provide seamless blending of light in runs of up to 50 feet. The LED fixtures reduce the electric load by 60% compared with 13-watt CFL cove lights. Their long useful life dramatically reduces the labor and maintenance costs of servicing lights installed up to 80 feet above the main floor.

areas where it may be difficult to install and maintain conventional light sources. LEDs also have no moving parts or filaments that can easily break down or fail.

- RGB and tunable white-light LED fixtures **can natively produce millions of colors or ranges of color temperature** without gels or filters.

- LED lighting systems can be **digitally controlled for maximum efficiency and flexibility**.

- LED lighting fixtures are **instant-on** — no warm-up or reset time, and they are unaffected by power cycling and dimming.

- Well-designed LED lighting systems offer **ease and flexibility of installation** — no ballasts, fewer power supplies, low-profile fixture housings, conventional wiring. Features such as current limiting and miswiring protection, power factor correction, and conduit-ready connections increase energy efficiency and ease of installation.

- Unlike fluorescent sources that contain mercury and require special handling and disposal, LEDs are **mercury-free and safe for the environment**.

- Many LED lighting fixtures are now **meeting or exceeding standards for energy efficiency and environmental friendliness**. Ongoing work on standards for testing and measurement will soon provide a basis for accurately comparing LED lighting fixtures with each other, as well as with conventional alternatives.

Greggs Plc, the UK's leading bakery retailer, lights the sales area of its concept store in Bromley, Kent entirely with LED lighting fixtures from Philips. The LED installation is expected to reduce energy consumption by 50% over conventional lighting, and eliminate 2 tons of carbon emissions per year. The fixtures' low heat output minimizes air conditioning needs for further energy savings during the summer.

2

LED Basics

AN LED IS a semiconductor device that emits visible light of a certain color. An LED is fundamentally different from conventional light sources such as incandescent, fluorescent, and gas-discharge lamps. An LED uses no gas or filament, it has no fragile glass bulb, and it has no failure-prone moving parts.

A Brief History of LEDs

LEDs, or *light-emitting diodes*, are electronic light sources. In 1962, the first red LED was developed by Nick Holonyak at General Electric. Throughout the 1960s, monochrome red LEDs were used for small indicator lights on electronic devices. Although they were dim and inefficient, the technology showed promise and quickly improved. Green and yellow LEDs were introduced in the early 1970s. These LEDs were used in watches, calculators, electronics, traffic lights, and exit signs. LEDs continued to improve in light output, and by 1990 LEDs of 1 lumen output were available in red, yellow, and green.

In 1993, Shuji Nakamura, an engineer at Nichia, created the first high-brightness blue LED. Because red, blue, and green are the three primary colors of light, LEDs could now produce virtually any color, including white. Phosphor white LEDs — LEDs that combine a blue or ultraviolet LED with a phosphor coating to produce white light — first appeared in 1996. By the late 1990s, LEDs began replacing incandescent sources in applications calling for colored light.

Between 2000 and 2005, output levels of 100 lumens and higher were achieved. White-light LEDs became available in various shades of light to match the warmth or coolness

> ✹ Lumens *are the standard measure of light output. A 60-watt incandescent bulb typically outputs a total of around 800 lumens. Lumens are discussed in detail in Chapter 3.*

15

of incandescents, fluorescents, and daylight. LEDs began competing with conventional light sources, and found their way into stage and entertainment lighting applications.

Today, LEDs represent viable sources for general illumination in most applications. The Department of Energy and the Optoelectronics Industry Development Association both expect LED technology to become the preferred method of lighting in homes and offices by 2025.

How LEDs Work

Like any diode, an LED consists of a single p-n semiconductor junction. Through a process known as *doping*, the n-type material is negatively charged and the p-type material is positively charged. The atoms in the n-type material have extra electrons, while the atoms in the p-type material have *electron holes* — electrons missing from their outer rings.

Applying a current to the diode pushes the atoms in the n-type and p-type materials toward the junction. When they get close enough to each other, the n-type atoms "donate" their extra electrons to the p-type atoms, which "accept" them. Applying a negative charge to the n-type side of a diode allows current to flow from the negatively-charged area to the positively-charged area. This is called *forward bias*.

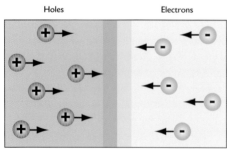

	LED Timeline
	First red LED developed by Nick Holonyak at GE in 1962
1960s	Red indicator LEDs manufactured by HP, with materials from Monsanto — .01 lumens
	First green and yellow LEDs
	First blue LEDs arrive in 1971
1970s	1 lumen red LEDs available by 1972
	LEDs used in watches, calculators, traffic lights, exit signs
1980s	Advances in lumen output
	1984: First superbright red LEDs
	High-brightness blue LEDs by Shuji Nakamura at Nichia in 1993
	1995: High-brightness green LEDs
	First white LEDs developed in 1996
	Ultrabright red and amber LEDs
1990s	LEDs begin to replace incandescent sources in colored light applications
	LEDs become viable for portable illumination applications
	Color Kinetics founded in 1997
	1998: RGB lighting applications
	White light via RGB LEDs
	White light via blue + phosphors
	First "tunable" white light LED fixtures
	LEDs available in 10 – 100 lumens
	By 2003, LEDs widely accepted in entertainment lighting applications
2000s	White-light LEDs become viable for accent lighting by 2004
	1000+ lumen LEDs via multichip packages available by 2005
	By 2008, LEDs become viable for general illumination
	Multiple manufacturers (Nichia, Cree, Osram, Lumileds, King Brite, Toyoda Gosei, Cotco . . .)

When extra electrons in the n-type material fall into the holes in the p-type material, they release energy in the form of *photons*, the basic units of electromagnetic radiation. All diodes release photons, but not all diodes emit light. The material in a light-emitting diode is selected so that the wavelength of the released photons falls within the visible portion of the light spectrum. Different materials produce photons at different wavelengths, which appear as light of different colors.

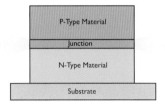

The beam of visible light from an LED is cool, but because they are not perfectly efficient, LEDs generate heat at the p-n junction — sometimes quite a lot of heat. Controlling the temperature at the junction with a well-designed heat sink and other thermal management features is critical for assuring proper operation, optimizing light output, and maximizing lifetime.

LED Anatomy

The two basic types of LEDs are *indicator-type LEDs* and *illuminator-type LEDs*. Indicator-type LEDs, such as 5 mm LEDs, are usually inexpensive, low-power LEDs suitable for use only as indicator lights in panel displays and electronic devices, or instrument illumination in cars and computers. Illuminator-type LEDs, also known as surface-mount LEDs (SMDs), high-brightness LEDs (HB-LEDs), or high-power LEDs (HP-LEDs), are durable, high-power devices capable of providing functional illumination with light output equal to or surpassing that of conventional light sources.

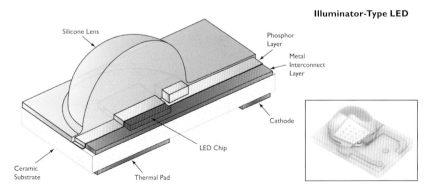

Illuminator-Type LED

Indicator-Type LED

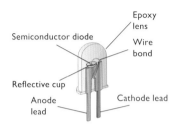

All illuminator-type LEDs share the same basic structure. They consist of a semiconductor chip (or *die*), a substrate that supports the die, contacts to apply power, bond wire to connect the contacts to the die, a heat sink, lens, and outer casing. (Some LEDs, like the TFFC from Philips Lumileds, require no bond wire.)

Since indicator-type LEDs are low power, any generated heat is dissipated internally. Illuminator-type LEDs, on the other hand, are packaged in surface-mount solder connections, and provide a thermally conductive path for extracting heat. This thermal path is critical for proper thermal management and operation of the LED.

How LEDs Produce Different Colors

LEDs produce different colors by using different *material systems*. Different material systems produce photons at different wavelengths, which appear as light of different colors.

The oldest LED technologies use materials such as gallium phosphide (GaP), aluminum gallium arsenide (AlGaAs), and gallium arsenide phosphide (GaAsP) to produce wavelengths from red to yellowish green. Nowadays, GaP, AlGaAs, and GaAsP are used almost exclusively in indicator-type LEDs, as the higher temperatures and electric currents required for illumination severely diminish the lifetimes of LEDs that use these materials.

Illuminator-type LEDs typically use newer material systems that can handle the necessary levels of current, heat, and humidity. High-brightness red and amber LEDs use the aluminum indium gallium phosphide (AlInGaP) material system. Blue, green and cyan LEDs use the indium gallium nitride (InGaN) system.

Together, AlInGaP and InGaN cover almost the entire spectrum, with a gap at green-yellow and yellow. Corporate colors that use yellow (such as Shell or McDonald's) are

The additive color model applies to light emitted directly from an illuminant. Combining red, green, and blue in equal amounts produces white.

The subtractive color model applies to reflective surfaces such as paints, dyes, and inks. Combining red, yellow, and blue in equal amounts produces black

therefore difficult to obtain with single-color LEDs. One method of achieving difficult-to-obtain colors is to mix different colors of LEDs in the same device.

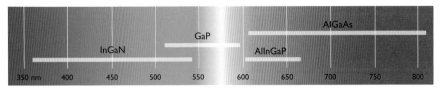

The main material systems for producing monochrome LEDs. AlInGaP and InGaN cover almost the entire spectrum for high-intensity LEDs, except for green-yellow and yellow at wavelengths between 550 nanometers (nm) and 585 nm. Colors in this gap can be achieved by mixing green and red LEDs.

How LEDs Produce Millions of Colors

LED manufacturers typically offer a variety of LED products in a range of colors — royal blue, blue, green, amber, red-orange, red, and so on. By itself, an LED can emit only the one color that the specific composition of its materials can produce. The real magic happens when LEDs of different colors are combined.

Combining red, green, and blue LEDs in a single LED device, such as a lighting fixture or multi-chip LED, and controlling their relative intensities can produce millions of colors. Like a television screen or computer monitor, a full-color LED device uses the *RGB color model* (r = red, g = green, b = blue). The RGB color model is an *additive color model*, which applies to light emitted directly from an illuminant. (The *subtractive color model* applies to reflective surfaces such as paints, dyes, and inks.)

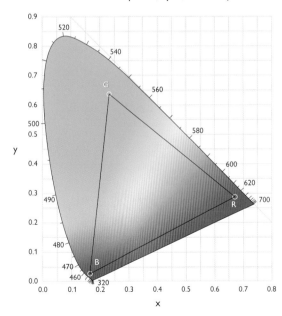

The diagram to the left shows the *CIE 1931 color space*, created by the International Commission on Illumination (CIE) in 1931 to define the entire range, or *gamut*, of colors visible to the average viewer. No single device — no TV screen, computer monitor, LED lighting fixture, or other tricolor device — has a gamut large enough to produce every color visible to the human eye. The gamut of colors that an LED fixture or multi-chip LED can produce depends on the specific colors of the individual red, green, and blue LEDs used in that fixture or chip.

The three color points of the individual LEDs used in a tricolor device describe a triangle. In theory, the device can reproduce every color point inside the triangle. In practice, since a tricolor LED device is usually controlled digitally, it can produce a *sampling* of the possible colors within the triangle. An 8-bit tricolor LED device can generate approximately 16.7 millions colors (256^3 colors) — but this is already more colors than the human eye can distinguish within the defined color triangle. (Colors outside the color triangle may be distinguishable, but the device can't produce them.)

The ability of full-color LED fixtures to produce virtually any color without gels, filters, or other extra equipment firmly differentiates LED lighting from any other type of lighting. Combining full-color LED lighting fixtures with lighting controllers can produce simple color-changing effects, intricate full-color light shows, and even large-scale video displays.

How LEDs Produce White Light

There are two methods of producing white light with LEDs:

- According to the RGB color model, white light is produced by the proper mixture of red, green, and blue light. The RGB white method produces white light by combining the output from red, green, and blue LEDs.

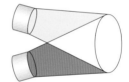

- The *phosphor white* method produces white light in a single LED by combining a short-wavelength LED, such as blue or UV, and a yellow phosphor coating. The blue or UV photons generated in the LED either travel through the phosphor layer without alteration, or they are converted into yellow photons in the phosphor layer. The combination of blue and yellow photons combine to generate white light.

White light can be produced by combining the wavelengths of yellow and blue light only. Sir Isaac Newton discovered this effect when performing color-matching experiments in the early 1700s.

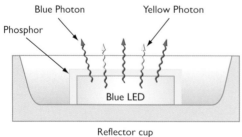

RGB white gives you control over the exact color of the light, and it tends to make colors "pop." But RGB white light is hardware-intensive, since it requires three LEDs, and it tends to render pastel colors unnaturally, a fact which is largely responsible for the poor *color rendering index* of RGB white light. (See "Appearance of Light" in Chapter 3 for a discussion of LEDs and color rendering.)

> ✹ Color rendering *is the ability of a light source to reproduce the colors of various objects faithfully in comparison with an ideal or natural light source.*

Phosphor white offers much better color rendering than RGB white, often on a par with fluorescent sources. Phosphor white light is also much more efficient than RGB white. Because of its superior efficiency and color rendering, phosphor white is the most commonly used method of producing white light with LEDs.

In a typical phosphor white manufacturing process, a phosphor coating is deposited on the LED die. The exact shade or color temperature of white light produced by the LED is determined by the dominant wavelength of the blue LED and the composition of the phosphor.

The thickness of the phosphor coating produces variations in the color temperature of the LED. Manufacturers attempt to minimize color variations by controlling the thickness and composition of the phosphor layer during manufacturing. Philips Lumileds, for example, uses a patented process to manufacture cool and neutral white Philips LUXEON LEDs with a high degree of color consistency.[10]

Tunable white light fixtures adapt the mixing principles of tricolor LED devices to produce white light in an adjustable range of color temperatures. Tunable white light devices typically combine cool and warm white LEDs, which can be individually controlled like the red, green, and blue LEDs in a full-color LED device. Adjusting the relative intensities of the warm and cool LEDs changes the color temperature in a tunable white device, just as adjusting the relative intensities of the red, green, and blue LEDs changes the color in a full-color device.

LED Fixture Anatomy

To be used for illumination, LEDs must be integrated into systems that incorporate optics, LED drivers, power supplies, and thermal management. Well-designed LED lighting fixtures integrate all of these critical components into the fixture itself.

As an LED lighting specifier or installer, you will likely never have a reason to open an LED lighting fixture's housing — just like, as a computer user, you likely never have a reason to open up your notebook computer's casing and dig around inside. Some theatrical fixtures are designed for replacing and repairing components in the field, but in most cases opening up an LED lighting fixture is inadvisable, is frowned upon by the fixture manufacturer, and may void the fixture's warranty. Fortunately, outright failure of LED fixtures and components is relatively rare. In the case of failure, manufacturers usually repair or replace fixtures for you under the terms of your warranty or other user agreement.

Nevertheless, it can be useful to have a general idea of what's inside the black box. The main components of a well-designed LED lighting fixture include:

- LEDs and supporting electronics
- Microprocessor-based power management, conversion, and control stages
- Thermal management features, such as vents and heat sinks
- Lensing and aiming features for directing, mixing, and dispersing light

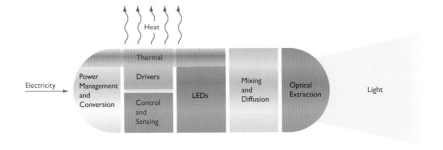

An LED lighting fixture is an integrated system consisting of LEDs, power management and conversion stages, LED drivers, control and sensing circuitry, thermal management features, and lensing and other optics for mixing, diffusing, and extracting light.

In addition to these common components, LED fixtures typically include integrated or detachable leader cables for connecting to power. Linear fixtures, such as cove lights and task lights, usually feature built-in fixture-to-fixture connectors, jumper cables, and other custom hardware for installing runs of fixtures with various spacing and geometries.

3

LED Lighting in Detail

LED LIGHTING FIXTURES are both similar to and different from conventional lighting fixtures. It's important to understand these similarities and differences in order to make accurate comparisons between conventional and LED-based lighting fixtures, and to specify the appropriate LED lighting fixture for a particular task or application.

As for the similarities, we've already seen that well-designed LED lighting fixtures offer features, performance, and economy conventional to lighting fixtures. Like conventional lighting fixtures, LED lighting fixtures are available in a variety of types, output levels, and sizes to support the full range of lighting applications. LED lighting fixtures can be installed and powered as easily as any conventional lighting fixture, using either conventional wiring or simple pre-configured cabling.

Conventional lighting fixtures typically consist of a lamp (a bulb, tube, or other light source) and a separate fixture housing.

An LED lighting fixture is a solid-state device integrating light sources (LEDs), housing, electronics, thermal management features, and lensing and aiming features.

One of the most obvious differences between LED lighting fixtures and conventional lighting fixtures is the fact that LEDs generate light using fundamentally different materials and means. A less obvious difference, but equally crucial to understanding how LED lighting fixtures differ from their conventional counterparts, is the fact that LED fixtures effectively erase the distinction between lamp and luminaire. In an LED fixture, the "lamps" — the LEDs themselves — are inseparable from the "luminaire" — the fixture's housing, electronics, and primary lensing.

These two differences have far-reaching consequences for the way LED lighting fixtures are tested, how their light output is measured and reported, how to accurately assess their suitability for a particular task or application, and how to make meaningful comparisons between LED lighting fixtures and conventional lighting fixtures.

With an understanding of the differences between LED lighting fixtures and their conventional counterparts, you can avoid some of the common pitfalls and confusions about LED lighting fixtures. With the ability to accurately interpret key specifications, you can effectively specify LED lighting fixtures for particular tasks and applications. The sections that follow offer a general overview of LED lighting technology, and pave the way for the material on LED lighting solutions and systems presented in Chapter 4.

Evaluating Light Output: The Importance of Delivered Light

Light output is an informal term for how much light a fixture produces, and how the fixture emits and distributes that light. The formal term for data describing the quantity and distribution of the visible light produced by a light source or fixture is *photometrics*.

It is common practice for lighting fixture manufacturers to report key photometrics in fixtures specification sheets or data sheets. These usually include basic charts and graphs that describe the power or intensity of a lamp or light fixture, the way it distributes light in space or over an area, and its energy efficiency. Lighting specifiers and designers use this data to make preliminary evaluations of a lighting fixture's capabilities and its potential suitability for a

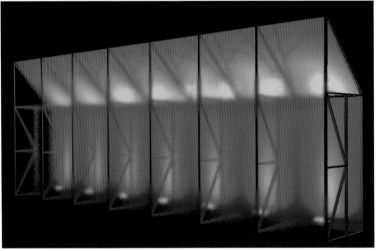

A computer graphic ray trace image, rendered in AGi32 lighting analysis software, from raw photometric data for ColorBlast Powercore LED fixtures from Philips Color Kinetics. The image models ColorBlast Powercore lighting effects in an exterior glass lightbox.

particular task or application. Fixture manufacturers often make more detailed information available for deeper analysis, creating computer renderings, building mock-ups, and so on.

The specification most commonly used for evaluating and comparing the performance of conventional lighting fixtures is *lumen output*. Technically speaking, the lumen is a standard unit that measures the total perceived power of a light source. (Exactly what "total perceived power of a light source" means is discussed in more detail below.) In practice, lighting professionals, purchasers, and users often refer casually to a fixture's "brightness." This is not only inaccurate, but it can be misleading, especially where LED lighting fixtures are concerned.

> ✺ Luminaire testing labs report raw photometric data in .IES files, the format of which conforms to standards set by the Illuminating Engineering Society of North America (IES). Reputable fixture manufacturers make these files available to lighting designers, who use them with lighting analysis software tools to create charts, graphs, and dimensional renderings, and to make comparisons between fixture types.

For a number of reasons, lumen output is not the best measurement of an LED lighting fixture's capabilities. In fact, evaluating an LED lighting fixture solely or primarily on the basis of its lumen output can underestimate or otherwise give a distorted picture of the fixture's performance and suitability for a given task or application.

Instead of lumen output, the best and most relevant measurement for evaluating LED lighting fixtures and for making accurate comparisons with conventional lighting fixtures is *delivered light*. The formal term for measurements of delivered light is *illuminance*. Roughly speaking, illuminance is the intensity of light falling on a surface area. If the area is measured in square feet, the unit of illuminance is *footcandles* (fc). If measured in square meters, the unit of illuminance is *lux* (lx).

Delivered light describes how much *useful light* a lighting fixture can deliver to a task area. Useful light is the portion of a lighting fixture's light output that is effectively directed to a task area, discounting any wasted light. The task area can be any space or surface that requires illumination — an entrance hallway, a common office space with desktop computers, a kitchen countertop, or the face of a Mayan pyramid in Guatemala. Light can be wasted in a number of ways: It can be partially blocked or dispersed within the fixture housing, it can be emitted in a direction away from the task area, or it can be lost through filtering, lensing, fixture positioning, or any of a number of other factors relevant to a specific installation.

A look through *The IESNA Lighting Handbook*, the 1,000-page standard reference work from the Illuminating Engineering Society of North America (IES),[11] demonstrates the importance of delivered light as a lighting principle, especially for white-light and everyday applications. Along with chapters that describe in detail how to deliver the right amount of useful light in an application, the

Clusters of eW Downlight Powercore, LED downlights from Philips Color Kinetics, and indirect lighting illuminate the dining room at Flinstering, a restaurant in Breda, The Netherlands. Recommended light levels for dining areas such as these is about 10 fc (100 lx).

Handbook includes an extensive lighting design guide that specifies ideal light levels for every conceivable interior, industrial, outdoor, sports, transportation, and emergency lighting application.

For example, the *Handbook* recommends a delivered light level of around 30 fc (300 lx) for an open plan office with extensive computer use, as well as for the ticket counter of a transportation terminal. Freight elevators should have a level of around 5 fc (50 lx), while serious reading in a chair at home requires a level of around 50 fc (500 lx). Lecture halls where demonstrations are being performed should have a light level of around 100 fc (1000 lx).[12]

> ✱ *Technically., there are 10.7 lux per footcandle. However, to simplify the math, lux measurements are typically given as footcandles x 10. 30 fc is therefore equivalent to 321 lx, but is expressed as 300 lx by convention.*

The Trouble with Lumens

The way that lumen output is traditionally measured, reported, and interpreted poses a number of potential problems for evaluating and comparing LED lighting fixtures:

- Since complete and accurate definitions of lumens and related photometric terms can be technical and complex, they are often misunderstood. Without a good understanding of these terms, however, the unique properties of LED lighting sources cannot be clearly grasped.

- Lumens are an imperfect measurement of the perceived intensity of light sources, with known shortcomings. The specific spectral properties of LED light sources exaggerate these shortcomings, especially toward the blue end of the spectrum.

- Conventional lighting fixture manufacturers often report total lamp lumens more prominently than or instead of total fixture lumens. Because many LED lighting fixtures do away with the distinction between lamp and luminaire, only total fixture lumens can serve as a basis for valid comparisons between LED and conventional lighting fixtures.

- LED lighting fixtures and conventional lighting fixtures are tested differently, and therefore some photometric data is reported differently. These differences must be taken into consideration in order to accurately compare conventional lighting fixtures and LED lighting fixtures.

- A fixture's total lumen output does not account for wasted light. Because LED lighting fixtures are fundamentally directional and natively create white and colored light without filtering or additional lensing and shading, LED fixtures typically waste much less light then their conventional counterparts, and deliver more of their total light output to a task or target area. An LED lighting fixture with lower rated lumens, therefore, may deliver the same or more useful light in a specific application than a comparable conventional lighting fixture with a higher rated lumen output.

Each of these issues is discussed in greater detail in the sections that follow.

What Exactly Is a Lumen?

Light measurements can either be *radiometric* or *photometric*. Radiometric measurements measure all the wavelengths of a light source, both visible and invisible. Photometric measurements measure only the visible wavelengths of light. The total electromagnetic energy that a light source emits across all wavelengths is known as *radiant flux*, and is measured in *watts*. The total energy that a light source emits across the visible wavelengths of light is known as *luminous flux*, and is measured in lumens.

Since visibility only has meaning in relation to a human viewer, photometric data takes into consideration the varying sensitivities of the human eye to different wavelengths (colors) of visible light. The sensitivity of a human

> ✺ In casual usage, the apparent power of a light source is often mistakenly referred to as a fixture's "brightness." Brightness is subjective, and varies depending on such factors as the distance of the light source from the viewer, the viewing angle, and the conditions of the light source's surroundings. Lumen measurements, in contrast, are based on carefully defined standards and test conditions, rather than on subjective impressions.

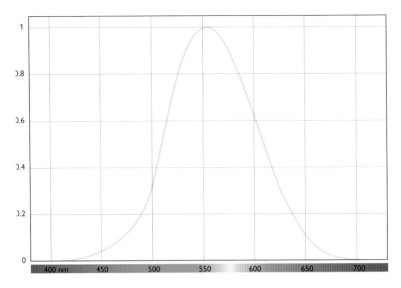

The spectral luminous efficiency function weights the perceived intensity of light of different wavelengths based on the varying sensitivities of the human eye. The eye is most sensitive to light around 550 nm in the green-yellow area of the spectrum, and is less sensitive at either the red or blue end.

eye with normal vision can be plotted as a bell-shaped curve. This curve is known as the *spectral luminous efficiency function*, and is often referred to as the *eye-sensitivity curve*. The eye-sensitivity curve shows that the human eye is most sensitive to light in the green part of the spectrum, around a wavelength of 550 nanometers (nm), and is progressively less sensitive to light toward both the red and blue ends of the spectrum.

To calculate lumens, different wavelengths of light are given more or less weight depending on where they fall on the eye-sensitivity curve. Two light sources with the same radiant flux falling on different parts of the curve will therefore have different lumen measurements. Imagine, for instance, two light sources of 1 watt of radiant flux each. One source emits a blue light at 480 nm, and one emits a green light at 555 nm. As the eye-sensitivity curve shows, the blue light appears significantly less bright than the green light, even though the total energy of the two lights is the same. To put it another way, the green light produces more lumens than the blue light, even though both lights produce the same amount of radiant energy.

> ✺ *Direct-view fixtures, such as tubes, panels, and string lights used for large-scale video displays, are designed for viewing rather than for illumination. The light output of direct-view fixtures is typically measured in candelas per square meter, sometimes called* nits. *Nits are a measure of* luminance, *the amount of light emitted or reflected from a particular area.*

In practice, there are variations in every individual's experience of the

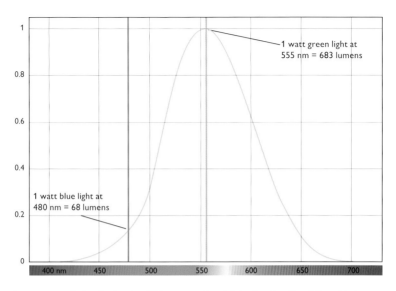

Lumens are calculated using the efficiency curve's weighting function. Two lights with the same radiant flux but producing light at different wavelengths, therefore, have different measured lumens.

apparent intensity of a light source. In 1924, the International Commission on Illumination (CIE), a recognized authority on light, illumination, color, and color spaces, standardized the responses of the human eye to visible light by defining a so-called *standard observer*. The standard observer has regular eye responses to visible light under specific conditions, which the standard defines. The eye-sensitivity curve used in lumens and other photometric measurements is the standard observer's eye-sensitivity curve, not the eye-sensitivity curve of any actual observer. Lumens and related measurements are therefore approximations or idealizations, which are usually good enough for evaluations and comparisons of different light sources.

Deficiencies of the Eye-Sensitivity Curve

It is well understood that the eye-sensitivity curve underestimates the perceived intensity of wavelengths toward the blue end of the spectrum. Although none has been widely adopted, various modifications of the eye-sensitivity curve have been suggested over the years. The Judd-Vos correction, for example, adjusts the curve to more accurately represent the normal sensitivity of human vision, especially to blue light.

The Judd-Vos correction, shown at the top of the next page, may not look like much, and it has relatively little effect when comparing conventional light sources with one another. But the correction can make a great deal of difference when measuring the luminous flux of LED light sources and comparing it to that of conventional light sources.

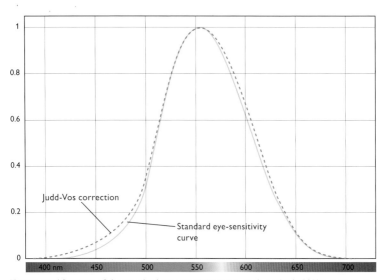

Various modifications of the standard eye-sensitivity curve, such as the Judd-Vos correction shown here, have been proposed to improve its accuracy in representing eye responses to different wavelengths of light.

Conventional light sources tend to radiate across a wide range of visible wavelengths. Incandescent light sources typically radiate throughout the visible band. Fluorescent light sources show a spiky spectrum, with intense radiation in some narrow wavelength bands and lesser levels of radiation throughout, due to emission lines of mercury, which LEDs do not contain. Solid-color LED light

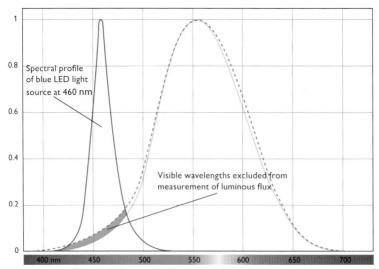

Visible wavelengths excluded from standard measurements of luminous flux can significantly underestimate the perceived intensity of some LEDs. The Judd-Vos modification partially corrects for this.

> ✱ *Saturated colors appear "brighter" to the eye than less saturated colors, even when their lumens are equivalent. This effect is not well understood as of yet, and is not well represented by the eye-sensitivity curve.*

sources usually radiate in a single, narrow band of wavelengths, which exaggerates discrepancies in the eye-sensitivity curve. The calculated lumens for a blue LED source with a peak wavelength of around 460 nm, for example, does not account for a significant portion of the visible light that the LED produces.

In practice, the deficiencies of the eye-sensitivity curve can result in lumen measurements that underestimate the perceived intensity of LED light sources, especially blue light. The perceived intensity of an LED lighting fixture, therefore, could be greater — in some cases much greater — than the fixture's reported lumens suggest.

Relative Photometry, Absolute Photometry, and Efficiency

Despite its shortcomings for accurately representing the perceived intensity of some LED light sources, luminous flux is a standard measurement used throughout the lighting industry. In evaluating lighting fixtures, you often have to compare the reported light output of LED lighting fixtures with the reported light output of conventional light sources. To make accurate comparisons, you need to understand the differences in the methods for gathering and reporting photometric data for conventional and LED lighting fixtures. With these differences in mind, you can avoid common pitfalls in interpreting and comparing the photometric data for LED and conventional fixtures.

The Labsphere LMS-200 20-inch Light Measurement Integrating Sphere.

Conventional lighting fixtures are tested using the *relative photometry* method. In relative photometry, luminaires and the lamps used within them are tested separately. Lamp and luminaire testing differ so much from one another, in fact, that lamp photometry engineering and luminaire photometry engineering are separate specialties, each with its own standards and practices. The total luminous flux and chromaticity (color) of a fixture's lamps are typically measured with *integrating spheres*, while the luminous intensity distribution and efficiency of the luminaires are usually measured with *goniophotometers* — although "gonies," as they are affectionately called, can also measure luminous flux.

In relative photometry, the lumen output of a fixture's lamps functions as a reference, and the lumen output of the luminaire is measured relative to it. A certain amount of the light from a conventional lighting fixture's lamps is typically blocked or absorbed by the fixture's housing, so total fixture lumens is always reported as a percentage of total lamp lumens. This percentage is the *efficiency* of the luminaire.

Because LEDs are typically inseparable from the luminaires in which they act as light sources, relative photometry is inappropriate for measuring the light output of LED fixtures. Instead, LED fixtures are tested using *absolute photometry*. The approved procedures and testing conditions for absolute photometry are spelled out in *Electrical and Photometric Measurements of Solid-State Lighting Products*, publication IES LM-79-08, published by IESNA in early 2008.[13]

In absolute photometry, only fixture lumens are measured, and not lamp lumens, because a separate measurement of the LEDs independent of the fixture is neither possible nor meaningful. Fixture efficiency, which compares lamp lumens to fixture lumens, therefore has no meaning for an LED lighting fixture. To put it another way, the efficiency of an LED lighting fixture in which the LEDs are inseparable components is always 100%.

Specifiers and designers sometimes mistakenly compare the total lamp lumens of a conventional lighting fixture with the total fixture lumens of an LED fixture. To make a valid comparison, you must reduce the measured lamp lumens of the conventional fixture by its efficiency. This reduction is typically reported in a Zonal Lumen Summary chart.

ZONAL LUMEN SUMMARY

Zone	Lumen	% Lamp	% Fixt.
0-30	113	12.1	19.6
0-40	199	23.1	34.6
0-60	382	44.4	66.3
0-90	534	62.1	92.8
90-120	38	4.4	92.8
90-130	40	4.7	7.0
90-150	41	4.8	7.2
90-180	41	4.8	7.2
0-180	575	66.9	100.0

The total fixture lumens of a conventional lighting fixture, which accounts for fixture efficiency, should be used in comparisons with the lumen output of LED fixtures.

For example, the Slique T2 SQ series of under-cabinet fluorescent light fixtures, from Alkco, prominently reports 860 lumens for the two T2 lamps used within the fixture. However, the Zonal Lumen Summary chart reports 575 total fixture lumens, because the fixture outputs only 66.9% of the total lamp lumens (66.9% of 860 = 575).[14] That means that 33.1% of the light produced by the fixture's lamps is wasted or lost within the fixture housing. When comparing a Slique T2 under-cabinet fixture to an LED-based under-cabinet fixture designed for similar lighting applications, you should compare the LED-based fixture's total lumens with the efficiency-corrected lumen total of the Slique T2, not with its reported lamp lumens.

Lensing, Filtering, Shading, and Other Sources of Loss

How much light a fixture delivers to a task surface depends on a complete range of factors in addition to the fixture's lumen output. These factors include fixture positioning, distance from the area to be illuminated, and light losses that result from lensing, filtering, shading, or other techniques or accessories used to direct or alter the source light.

The fact that LED lighting fixtures are inherently directional minimizes losses associated with lensing and shading. Furthermore, the fact that LEDs natively generate colored light eliminates losses associated with filtering used to alter the color or distribution of light produced by conventional lighting fixtures.

Filtering can block a significant percentage of a fixture's total lumens. Some blue and red filters can block 96% or more of a conventional floodlight's light output.[15] LED floodlights such as ColorReach™ Powercore, shown here illuminating the Royal Military Academy in Breda, The Netherlands, can natively produce intense, saturated color without filters.

General Lighting Example: The Downlight

A client has asked you to specify lighting for the corridors in a new office space. The client wants you to investigate LED-based alternatives to CFL downlights for energy-efficiency and maintenance reasons, but he isn't convinced that LED downlights are bright enough for the task.

You know that brightness is a subjective impression, and that the client is actually referring to luminous flux. When you compare the lumen output of some typical LED downlights with similar incandescent and CFL downlights, the LED downlights seem to be less capable — that is, they generally report fewer total lumens than their non-LED counterparts. But you also know that you need to correct for efficiency and other losses, and that the crucial measurement for the lighting application is not how "bright" the fixture is, but how much useful light it can deliver to a task area at a certain distance.

According to *The IESNA Lighting Handbook*, the ideal light level for corridors, lobbies, and other common areas in an office space is 5 – 10 fc (50 – 100 lx). The task area for this sort of illumination is a plane 30 in (0.75 m) above the floor. Any ceiling-mounted downlight that can deliver up to 10 fc (100 lx) to this task surface is appropriate for the application.[16]

eW Downlight Powercore is a surface-mounted LED downlight from Philips Color Kinetics. According to published specs, eW Downlight Powercore generates between 405 and 527 total fixture lumens, depending on color temperature and beam width. A comparable CFL downlight produces a total of 860 lamp lumens with two 13-watt T4 lamps. Information published by the National Lighting Product Information Program (NLPIP), an independent lighting research center providing objective test information to lighting professionals, gives this particular CFL downlight an efficiency rating of 50.1.[17] This means the CFL downlight actually outputs 430 total fixture lumens — on a par with eW Downlight Powercore.

The NLPIP created a corridor installation mock-up to measure the light levels delivered by the CFL downlight and other comparable downlights. Their test results indicate that, at a 9 ft ceiling height, the CFL downlight delivers an average of 11 fc (110 lx) to the task surface. According to published specs, eW Downlight Powercore fixtures can deliver an average of 15 fc (150 lx) at 9 ft, using a narrow beam angle. eW Downlight Powercore, therefore, delivers more light in this application than the CFL downlight. Not only that, the LED downlight's useful life is 10 to 20 times longer than the CFL downlight's, and it uses 40% less energy to operate (15 watts as opposed to 26 watts).

A mock-up of an architect's studio, showing LED downlights illuminating drafting desktops, the target work surface in this application. Performance of visual tasks in this environment requires a light level of about 50 fc (500 lx) at the work surface.

Cove Lighting Example: Directional Light

LED lighting fixtures with integrated optics and lensing can direct light to a target application area more efficiently than fluorescent and standard bulb-shaped incandescent lamps, which emit light in all directions. A significant percentage of the light produced by a typical fluorescent or incandescent lamp is lost within or blocked by the luminaire, reabsorbed by the lamp, or emitted in a direction

An LED cove light with integrated lensing emits 100% of its 177 fixture lumens within a 110° beam angle. A fluorescent cove fixture emits 85% of its 700 lamp lumens in all directions, delivering 182 lumens in any 110° slice — effectively the same amount of delivered light as the LED cove light.

that is not useful for the intended application. For some fixture types, such as recessed downlights, troffers, and under-cabinet fixtures, 40% to 50% of the total lamp output is lost before it exits the fixture.

LED fixtures that emit light in a specific direction reduce the need for reflectors and diffusers that can trap light, and can therefore deliver light more efficiently to a target area. For example, eW Cove Powercore, a linear LED fixture from Philips Color Kinetics, emits light in a tight 110° spread. At 177 lumens per foot, these fixtures produce much less light than a popular F32T8 lamp, which is rated at 700 lumens per foot. However, analysis shows that eW Cove Powercore delivers a comparable level of light to a target area.

Including all losses, about 85% of the F32T8's lamp lumens leave the fixture, reducing its lumen output to 595 lumens per foot. But those 595 lumens are radiated in all directions, in 360°. Any 110° slice would contain about 30% of the lumen output, or 182 lumens — almost exactly the same as eW Cove Powercore. Because it contains an integrated lens and housing that rotates through a full 180°, eW Cove Powercore lets designers and installers easily direct light exactly where it's needed, without secondary lensing or diffusers that would almost certainly reduce the fixture's light output.

Quality of Light

Quality of light is a concern for both colored light and white light. While color consistency, saturation, and accuracy concern colored and white light equally, white light for general illumination involves additional concerns. Two key measures of the quality of white light are *correlated color temperature* (CCT) and *color-rendering index* (CRI). Correlated color temperature describes whether white light appears warm (reddish), neutral, or cool (bluish). Color-rendering index describes how well a light source renders colors.

White-light LEDs are now capable of producing color temperatures both more consistently and in a wider range than almost any other light source.

White-light LEDs are also approaching and in many cases surpassing conventional sources in their ability to render colors accurately.

Color-Rendering Index and White-Light LEDs

Color rendering index measures the ability of a light source to reproduce the colors of various objects faithfully in reference to an ideal light source. The index rates the quality of a light source on a scale up to 100. By definition, the reference source — the sun, for example, or an incandescent light source — has a CRI of 100. The best possible faithfulness to the reference source also has a CRI of 100.

The CIE has developed a test that measures how much eight standardized color samples, designated R1 through R8, differ in appearance when illuminated under a given light source, relative to the reference source. These eight sample colors have a relatively low saturation, and are evenly distributed over the complete range of hues. Some lighting manufacturers also use an R9 sample, a saturated deep red color. The color rendering score, usually notated as R_a, is derived from the results for all sample colors tested.

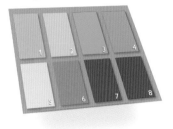

Eight standard color samples are used in the conventional method for measuring CRI.

Do LED Light Sources Produce Acceptable CRI?

Minimum acceptable CRI for a light source depends on the application:

- A CRI of 90 – 100 is required in retail and work spaces where faithful color rendering is crucial — for instance, in shops displaying fine merchandise or artwork, or in graphic design studios.

- Most office, retail, school, educational, medical, and other work and residential spaces require a minimum CRI of 70 – 90.

- Industrial, security, and storage lighting where color fidelity is not important can use sources with minimum CRI as low as 50.

Lighting fixtures using phosphor white LEDs are available today with CRIs of 80 or better, comparable to typical CFL lamps, quartz metal halide lamps, and some cool white fluorescent fixtures. LED lighting fixtures that use these LEDs deliver CRIs appropriate for the vast majority of applications.

Shortcomings of CRI for White-Light LEDs

CRI has been used to compare incandescent, fluorescent, and HID lamps for many years, but the CIE has concluded that CRI cannot effectively predict the color quality of white-light LEDs (CIE Technical Report 177:2007, *Color Rendering of White LED Light Sources*).[18]

The CIE's conclusions are based on numerous academic studies and experiments demonstrating that observers tend to rank the quality of scenes illuminated with LEDs much higher than their calculated CRI values would suggest. Some phosphor white LEDs and RGB white LED clusters receive CRI scores as low as 20. Nevertheless, observers consistently rate LED white light as visually appealing.

The explanation for these discrepancies is fairly technical, but the bottom line is that standard CRI tests favor light sources that mimic either *black-body radiators* — solid objects that emit particular colors of light when heated, such as incandescent lamp filaments — or daylight. The phosphor coating of many fluorescents have been tuned over the years to achieve good CRI scores, but these changes have had little effect on the observed color fidelity of fluorescent sources.

Researchers at the National Institute of Standards and Technology (NIST) are in the process of developing a Color Quality Scale (CQS) that would do a better job than CRI of measuring the color-rendering abilities of all white light sources, including white-light LEDs. According to NIST, CQS evaluates various aspects of color quality, including color rendering, chromatic discrimination, and observer preferences. In addition to several other recommendations, CQS replaces the eight color samples used in CRI calculations with 15 color samples that more completely represent the range of normal object colors, and which take the spectral properties of LEDs into account.[19]

> ❋ Please visit the Department of Energy's Solid-State Lighting pages at www1.eere.energy.gov/buildings/ssl/index.html for a detailed technical discussion of CRI testing and LEDs.

CQS is still under development, and has not yet been widely adopted by luminaire testing labs, which continue to use standard CRI testing despite its known shortcomings. Until CQS or a similar alternative is in place, specifiers and designers should observe LED sources with low CRI scores in person — if possible, on site — to evaluate how well they render color.

LEDs and Color Consistency

Color consistency is an index of light quality for both color and white-light LEDs. Where white light is concerned, *correlated color temperature, or* CCT, describes whether white light appears warm (reddish), neutral, or cool (bluish). The standard definitions of CCT allow a range of variation in chromaticity that can be readily discerned by viewers even when the CCT value is the same. Ensuring color consistency, therefore, is a major concern of LED manufacturers, who devise methods to keep color variations under tight control.

Understanding Correlated Color Temperature

Technically speaking, the "temperature" in correlated color temperature refers to *black-body radiation* — the light emitted by a solid object with certain properties, heated to the point of incandescence — and is expressed in degrees K (Kelvin), a standard measurement of absolute temperature. As a black body gets hotter, the light it emits progresses through a sequence of colors, from red to orange to yellow to white to blue. This is very similar to what happens to a piece of iron heated in a blacksmith's forge. The sequence of colors describes a curve within a color space, as the following diagram of the CIE 1931 color space shows.

> ✺ A *black body* is an object that absorbs all electromagnetic radiation falling on it. Because it reflects no light, a black body appears black. No perfect black bodies exist, but certain metals offer good approximations.

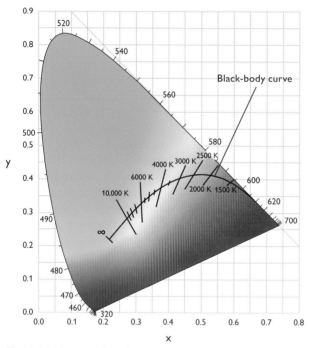

The black-body curve defines the range of color temperatures, from warm (reddish) to cool (bluish), within the CIE 1931 color space.

An incandescent lamp emits light with a color of roughly 2700 K, toward the warm or reddish end of the scale. Because an incandescent bulb uses a filament that is heated until it emits light, the temperature of the filament is also the color temperature of the light.

Spectral analysis of visible light makes it possible to define color temperatures for non-incandescent white light sources, such as fluorescent tubes and LEDs. The actual temperature of a 2700 K LED is typically around 80° C, even though the LED emits light of the same color as a filament heated to a temperature of 2700 K.

As the diagram opposite shows, any light source with a measured chromaticity that falls along one of the color temperature lines, which lie perpendicular to the black-body curve, has that color temperature, even through light sources with the same defined CCT can display fairly large differences in hue. For this and related reasons, LED makers use a method of managing manufacturing variations in chromaticity (and other characteristics) known as *binning*.

Consistent, Reliable Color: All About Binning

During production, LEDs vary in color, luminous flux, and forward voltage. Since the differences are significant, LEDs are measured and delivered to the market in subclasses, or *bins*. Binning makes it possible to select LEDs that conform to stated specifications — for instance, to select LEDs for traffic signals with the specific color required to meet the European standard.

One important goal for lighting fixture manufacturers is to select bins of LEDs in such a way as to minimize differences in color that might be visible from fixture to fixture or from production run to production run.

To understand how a bin is defined, we return to the diagram of the CIE 1931 color space, and zoom in on the black-body curve. Changes in color temperature lie along the black-body curve, but LED color variations also occur above and below the black-body curve. LEDs with color points above the black-body curve are greenish in tint, while those below the black-body curve are pinkish. In practice, this means that specifying a color temperature does not ensure color uniformity. For example, the two charts below illustrate two hypothetical LED bins, both centered on a color temperature of 5300 K, with a variation of + / - 300 K. Bin 1 has some color variation, since it extends above and below the black-body curve. Bin 2 has about four times as much color variation, even though it also conforms to the manufacturer's published specification for color temperature.

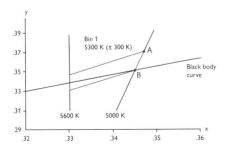
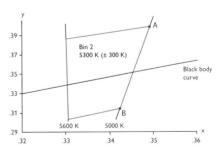

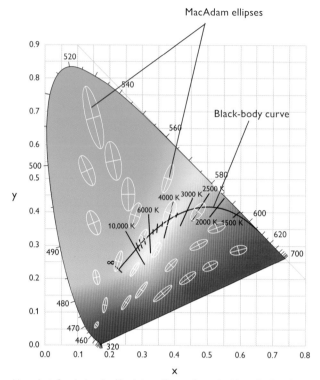

Note that, for clarity, the MacAdam ellipses shown in this and other diagrams are 10 times larger than actual size.

The threshold at which a color difference becomes perceptible is defined by a *MacAdam ellipse*. A MacAdam ellipse is drawn over a color space such that the color at its center point deviates by a certain amount from colors at any point along its edge. The scale of a MacAdam ellipse is determined by the *standard deviation of color matching* (SDCM). A color difference of 1 SDCM "step" is not visible; 2 to 4 steps is barely visible; and 5 or more steps is readily noticeable. As the illustration above shows, the size and orientation of MacAdam ellipses differ depending on their position within a color space, even when each ellipse defines the same degree of deviation between the color at its center and a color along its edge.

ANSI Chromaticity Standard C78.377A defines 8 nominal CCTs, each with color ranges defined by boxes surrounding 7-step MacAdam ellipses.[20] LEDs that fall within both the defined CCT and color ranges, therefore, conform to the standard.

ANSI C78.377A CCT Standard	
Nominal CCT	CCT Range (K)
2700 K	2725 ± 145
3000 K	3045 ± 175
3500 K	3465 ± 245
4000 K	3985 ± 275
4500 K	4503 ± 243
5000 K	5028 ± 283
5700 K	5665 ± 355
6500 K	6530 ± 510

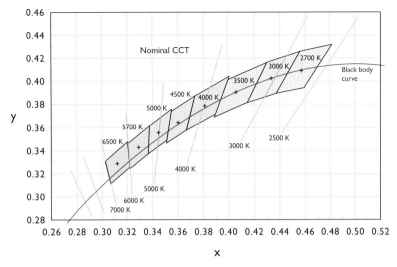

Color differences within the areas that conform to the CCT and chromaticity standards are readily perceptible. In practice, therefore, LED manufacturers subdivide each area into multiple bins. OSRAM, for example sells a certain number of bins for a color temperature, each of which falls within the area that conforms to the ANSI standard for that chromaticity. The diagram below shows an example binning plan for OSRAM Golden DRAGON LEDs at 2700 K.[21]

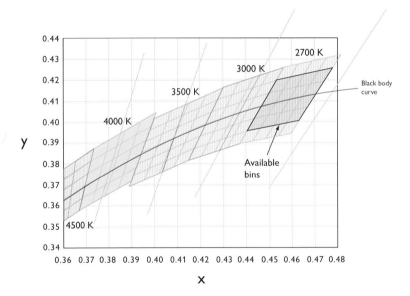

Even though all 16 bins that OSRAM offers conform to ANSI C78.377A for a nominal CCT of 2700 K, there will be variation in CCT and hue from bin to bin.

ANSI C78.377A goes part of the way toward ensuring color consistency, but some LED lighting manufacturers are adopting standards for LED purchase and use that exceed ANSI C78.377A. For instance, Philips Color Kinetics has developed a mathematical model for sorting, called Optibin®, that guarantees color uniformity across fixtures and production runs.

Optibin's CCT and hue tolerances for LED fixtures fall within a 4-step MacAdam ellipse, rather than within the 7-step ellipse defined by the ANSI standard. To ensure that variations in color will be barely noticeable, Optibin dictates the use of LEDs from bins which lie as close as possible to the black-body curve within the 4-step ellipse. A proprietary algorithm intelligently combines LEDs from different bins for each fixture production run, guaranteeing color uniformity for fixtures purchased and shipped at different times.

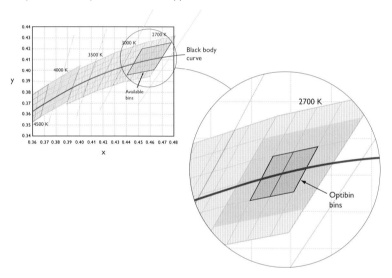

Choosing the Right White

LEDs can be made in a wide range of color temperatures that approximate the color temperatures of many non-LED lighting sources, daylight, and skylight. To select the right color temperature for an application, you must consider a number of factors.

Particular color temperatures, from warm to neutral to cool, are associated with certain light sources and environments. Color temperature also alters the emotional effect of a space, and can dramatically affect the appearance of objects on display in stores, galleries, and museums. Selecting the right color temperature matches light source to environment, and can positively influence buyer behavior and increase productivity in the workplace.

White-light LED fixtures with fixed color temperatures can easily replace most conventional light sources. Tunable white-light LED fixtures offer a range of color temperatures that can be varied on the fly with lighting controllers.

Warm Color Temperature

Cool Color Temperature

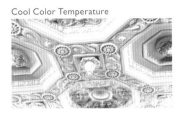

Tunable white light is ideal for illuminating changing retail displays, for altering the mood of a public space (with different morning, evening, and overnight lighting schemes, for example), and for theatrical or studio applications that call for varying levels and shades of white light.

	Effect, Mood, and Application by Color Temperature				
Color Temperature	Warm 2700 K	White 3000 K	Neutral 3500 K	Cool 4100 K	Daylight 5000 K – 6500 K
Effects and Moods	Warm Cozy Open	Friendly Intimate Personal Exclusive	Friendly Inviting Non-threatening	Neat Clean Efficient	Bright Alert Exacting coloration
Applications	Restaurants Hotel lobbies Boutiques Homes	Libraries Office areas Retail stores	Showrooms Bookstores Office areas	Office areas Classrooms Mass merchandisers Hospitals	Galleries Museums Jewelry stores Medical exam areas

Delivering the Whole Range of Color Temperatures

Like fluorescent sources, LEDs can deliver the whole range of color temperatures from warm to neutral to cool to daylight. Although mixing light source types in a single installation is not usually recommended, LEDs can replace fluorescent, halogen, incandescent, or metal halide sources in retrofit applications.

Tunable white light fixtures, such as the line of IntelliWhite® fixtures from Philips Color Kinetics, offer a range of color temperatures in a single fixture.

	Light Sources and Their Color Temperatures				
Color Temperature	Fluorescent	Halogen	Incandescent	LED	Metal Halide
Warm 2700 K	✓		✓	✓	
White 3000 K	✓	✓		✓	✓
Normal 3500 K	✓			✓	
Cool 4100 K	✓			✓	✓
Daylight 5000 K – 6500 K	✓			✓	

Efficacy of LED Lighting Fixtures

The *efficacy*, or energy efficiency, of lighting fixtures is the amount of light produced (in lumens) per unit of energy consumed (in watts), or lm / W.

Generally speaking, LEDs with the highest efficacy are the reds and the coolest whites (most bluish) — 5000 K and above. As of late 2009, LEDs far surpass incandescent lamps in efficacy, and compare favorably to most fluorescent lamps. Warm white LEDs in the 2600 K to 3500 K range approach the efficacy of CFLs, and continue to improve. Efficacies of 150 lm / W have been achieved in the lab, but the best production LEDs produce around 100 lm / W. The highest LED fixture efficacies are currently in the 50 lm / W range or higher.

> ❇ **Efficacy vs Efficiency**
> *Efficacy and efficiency are two separate measures of fixture performance, although they are often confused with one another. Efficiency is the ratio of lamp lumens to fixture lumens for a conventional lighting fixture. Efficacy is light output in lumens per unit of input energy in watts. It helps to remember that efficacy measures how much light a fixture produces per unit of electrical energy consumed, and efficiency measures how much of the total light produced by a fixture is wasted.*

Comparing the Efficacy of LED and Conventional Lighting Fixtures

To accurately compare the efficacy of an LED lighting fixture and a conventional lighting fixture, you must compare the efficacy of each as a whole system, including light source, power supply, ballast, electronics, fixture housing, and optics. When incorporated into a lighting fixture, the efficacy of both LEDs and conventional lamps is significantly reduced — and for many of the same reasons.

Fluorescent and HID light sources require ballasts to provide starting voltage and to limit electrical current to the lamp. LEDs similarly require drivers and other electronics to convert line voltage to a voltage the LEDs can use, to control electrical current, and to enable dimming and color correction. In general, LED drivers are about 85% efficient. For this reason alone, the efficacy of an LED reported by the manufacturer must be discounted by about 15% when incorporated in a fixture.[22] Lensing, operating temperatures, and other factors further discount reported efficacy of LEDs. The efficacy of non-LED lamps must be similarly discounted to account for light losses due to fixture housing, lensing, filtering, and so on.

Some Real-World Examples

The lamp efficacy for a recessed CFL downlight is 72 lm / W, but losses when incorporated in a luminaire account for an efficacy reduction of 67%, resulting in a fixture efficacy of 24 lm / W. The efficacy of a well-designed LED downlight is only 66 lm / W, but losses account for an efficacy reduction of 49%, resulting in a fixture efficacy of 34 lm / W — better than the conventional downlight.

Similarly, the rated lamp lumens for a 4000 K CFL wall-washing fixture is 85 – 90 lm / W, but luminaire losses account for 62% of that total, putting fixture efficacy at around 34 lm / W. The CFL fixture is therefore comparable in efficacy to a 4000 K LED wall washing fixture at around 35 lm / W.

The table below shows the starting efficacy of the lamps or light sources for various types of lighting fixtures, the total percentage of light losses resulting from the fixture, ballasts, drivers, lensing, and so on, and the net efficacy. The surface-mounted LED downlight (eW Downlight Powercore from Philips Color Kinetics is used in this example) has the lowest starting efficacy, but its net efficacy exceeds that of the asymmetric linear fluorescent fixture and equals that of the recessed CFL downlight.

Fixture	Starting Efficacy	Losses	Net Efficacy
Asymmetric TL Wall Wash	80 lm / W	62%	30 lm / W
Recessed CFL Downlight	72 lm / W	47%	34 lm / W
Surface-Mounted LED Downlight	66 lm / W	49%	34 lm / W
Indirect TL Troffer	68 lm / W	25%	51 lm / W
Direct / Indirect TL Pendant	74 lm / W	20%	59 lm / W

Source: US Department of Energy

Minimizing Off-State Power Consumption

One factor that is often overlooked but that can significantly reduce system efficacy is off-state power consumption. Off-state power consumption occurs when fixture switches or dimmers are positioned between the power supply or transformer and the fixtures. In configurations like this, the transformer continues to consume power even when the fixtures are turned off. Off-state transformer power draw can sometimes exceed 2 W,[23] and the resulting total losses can represent as much as 20% of a system's total load.

Off-state power consumption is eliminated in some LED lighting fixtures that can be powered directly from line voltage. One example of such a fixture is eW Profile Powercore from Philips Color Kinetics, where the power stage is integrated into the fixture's electronics and is therefore "downstream" from the system's switches or dimmers.

The Importance of Thermal Management

A common misconception about LEDs is that they generate no heat. Although LEDs do not *radiate* heat — they produce a cool beam of light — they do indeed generate heat.

LEDs convert electric power into radiant energy plus heat, like any light source. The proportion of heat to radiant energy is a function of input power and system efficiency. Incandescent lamps generate a high percentage of infrared energy (IR) and heat, and only a small percentage of visible light. Fluorescents and metal halides produce a higher proportion of visible light, but they also generate IR, ultraviolet (UV), and heat. Surprisingly, LEDs convert a relatively low percentage of electric power into radiant energy — about the same as fluorescent and metal halide sources — but since they radiate negligible amounts of IR and UV, the percentage of visible light that they produce is comparable to fluorescents and metal halides, and higher than incandescents.

The following table compares the percentage of input power converted to radiant energy and heat by LEDs to the percentages for some traditional light sources. For comparison purposes, these figures reflect output for white-light sources.[24]

Power Conversion Percentages for White-Light LEDs and Traditional Sources				
	LED	Incandescent	Fluorescent	Metal Halide
Visible Light	15% – 25%	8%	21%	27%
IR	~0%	73%	37%	17%
UV	0%	0%	0%	19%
Heat	75% – 85%	19%	42%	37%

Source: US Department of Energy

Effective thermal management is critical to LED fixture performance, as excess heat reduces light output and shortens the useful life of LED sources. For proper operation and maintenance, generated heat must be conducted away from the sources. Well-designed fixtures use effective heat sink designs and other conduction and convection features to channel heat away from the LED sources and dissipate it into the fixture's surroundings.

About Junction Temperature

One of the most important thermal measurements of LED operation is *junction temperature*, usually notated as T_j. As described in Chapter 2, the junction is the location in the LED where input energy (electrons) is converted into visible light (photons) plus heat. As junction temperature increases, light output and LED lifetime decrease.

Three factors affect the junction temperature of an LED: drive current, thermal path, and ambient temperature. In general, the higher the drive current, the greater the junction temperature. The amount of heat that can be conducted or convected away depends on the ambient temperature and the design of the thermal path from the LED to the fixture's surroundings.

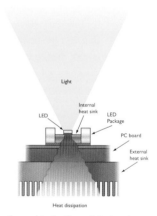

The thermal path of a well-designed LED lighting fixture. Heat sinks and other conduction and convection features channel heat away from the LEDs.

Typical high-power LEDs include an emitter, a circuit board, and a heat sink. The emitter is soldered onto the circuit board and contains the die, optics, and heat sink slug, which draws heat away from the die. Most LED fixtures use a metal-core printed circuit board (MCPCB), which marries a standard circuit board to a sheet of aluminum. The MCPCB transfers heat from the LED's heat sink to the external heat sink, on which the circuit board is mounted. The external heat sink — a device integrated into the fixture, or the fixture housing itself — in turn transfers the heat to the surroundings. If an LED lighting fixture had no external heat sink, or if the heat sink became blocked or covered, the LEDs inside the fixture would be destroyed within minutes.

How Junction Temperature Affects Light Output

LED manufacturers measure the luminous flux of their LEDs using a 15 – 20 millisecond power pulse, at a fixed junction temperature of 25° C. Junction temperatures of well-designed LED fixtures under standard operation — constant power at room temperature with heat convection and conduction features in place — typically range from 60° C to 90° C, or even greater. Because operational junction temperatures are almost always higher than 25° C, in-fixture LEDs produce at least 10% less light than the LED manufacturer's rating, unless the manufacturer supplies data for higher temperatures.

> ✻ *Increasing junction temperatures can cause LEDs to color-shift. This shift is usually visible only with amber LEDs, which are the most sensitive to temperature changes. Although the shift is small, it's important to check when color output must conform with standards — for example, for traffic signals.*

The following chart shows how LEDs of various colors can respond to operation at increasing junction temperatures. Amber and red LEDs are the most sensitive to changes in junction temperature, while blue is the least sensitive.[25]

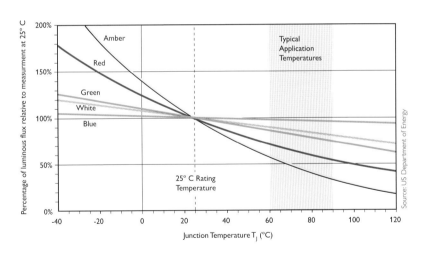

How Junction Temperature Affects Useful Life

Continuous operation at high junction temperatures dramatically shortens the useful life of LED lighting fixtures. The following chart shows light output over time, based on measurements to 10,000 hours and projections to 100,000 hours, for two identical LEDs driven at the same current but with different junction temperatures. With an increase of 11° C in junction temperature, estimated useful life decreases 57%, from around 37,000 hours to around 16,000 hours.[26]

> ✹ *Well-designed LED lighting fixtures incorporate thermal monitoring and thermal protection circuitry, which dims or turns off the fixture if it gets too hot. Auto-cycling features restore normal operation once junction temperatures drop to a safe level, or after a predetermined timeout.*

LED makers continue to improve the durability of LEDs at higher operating temperatures. For example, the long-term lumen maintenance application note for the Cree XLamp XR-E White LED, published in July 2009, claims 70% lumen maintenance to 50,000+ hours at a drive current of 700 mA, junction temperature at 110° C, and ambient temperature at 45° C.[27]

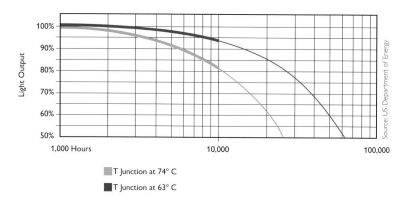

■ T Junction at 74° C
■ T Junction at 63° C

Useful Life: Understanding LM-80, Lumen Maintenance, and LED Fixture Lifetime

As with key photometric measurements, such as lumen output and efficacy, life calculations for LED light sources and conventional light sources differ considerably. Reliable comparisons between conventional and LED light sources require an understanding of these differences, and often involve some analysis of reported figures.

While life testing might appear to be as straightforward as turning a unit on and seeing how long it lasts, measurement and evaluation are not so simple, especially for LED light sources. Today, life testing methods for conventional light sources — incandescents, fluorescents, high-intensity discharge lamps, low-pressure sodium lamps, and so on — are well established and well understood. The life testing method for LED light sources is relatively new and less well understood.

This section explains how to interpret useful life calculations for LED sources and LEDs sources incorporated into lighting fixtures, and suggests a method for making accurate comparisons between conventional lamps and LED lighting fixtures.

Rated Lamp Life of Conventional Sources

Approved methods for life testing of conventional light sources call for measuring and reporting *rated lamp life*. These methods are published by the IES in a variety of official publications. For example, LM-65-01 defines life testing procedures for compact fluorescent lamps (CFLs), while LM-49-01 defines life testing procedures for incandescent filament lamps. Both LM-65-01 and LM-49-01 have been available and in use since 2001, and both methods revise older standards, published in 1991 and 1994 respectively.

> ❂ Life calculations for conventional light sources are typically expressed as rated lamp life, the mean time to failure of a statistically valid sample of lamps.

Both publications establish testing conditions, sample sizes, and methodologies for generalizing test data to arrive at rated life specifications. For CFLs, LM-65 specifies a statistically valid sample to be tested at an ambient temperature of 25° C, in a cycle of three hours on and 20 minutes off (as CFL life is appreciably shortened by the frequency with which the lamp is started). The point at which half the lamps fail is the rated average life for that lamp.

For incandescent filament lamps, LM-49 specifies a statistically valid sample to be tested within the manufacturer's stated operating temperature range and voltage. Lamps are allowed to cool to ambient temperature once a day (usually for 15 to 30 minutes). As with CFLs, rated life for incandescent filament lamps is the total operating time at which half the lamps are still operating.

Lumen Maintenance and Lumen Depreciation

In September 2008, the IES published *Measuring Lumen Maintenance of LED Light Sources*, publication IES LM-80-08. LM-80 is the LED counterpart of LM-65, LM-49, and other life testing standards for conventional light sources, but it differs from the older standards in important — and potentially confusing — ways.

> ❂ Lumen maintenance is the industry-standard term for the percentage of initial lumens that a light source maintains over time.

Instead of measuring rated lamp life, LM-80 calls for measuring how much an LED source's lumen output decreases over a certain number of hours of operation. Technically, the term for this decrease is *lumen depreciation*.

World Market Center in Las Vegas, Nevada, uses over 8,000 ft (2,438.4 m) of eW Cove Powercore linear LED lighting fixtures from Philips Color Kinetics to illuminate the complex geometries of the atrium in its Building C. LED cove lighting offers advantages over conventional cove lights in high-end installations such as this one. With useful life of up to 70,000 hours, and a very low failure rate, LED cove lights virtually guarantee reliable illumination around the clock for many years, without dark spots from lamp outages that can mar the elegant and uniform presentation of a premier space.

The converse of lumen depreciation is *lumen maintenance*, the industry-standard term for the percentage of initial lumens that a light source maintains over a certain period of time.

All electric light sources lose lumen output over time — indeed, annexes to both LM-65 and LM-49 address lumen depreciation of CFLs and incandescent filament lamps. In incandescent lamps, lumen depreciation is caused by depletion of the filament and the build-up of evaporated tungsten particles inside the bulb. Incandescents typically lose 10% – 15% of their initial lumen output over an average lifetime of 1,000 hours. In fluorescent lamps, lumen depreciation is caused by photochemical degradation of the phosphor coating and glass tube, and the build-up of light-absorbing deposits inside the tube. High-quality fluorescent lamps using rare earth phosphors lose only 5% – 10% of initial lumens over 20,000 hours of operation. CFLs depreciate more, but the most well-designed products lose no more than 20% of their initial lumens over an average lifetime of 10,000 hours.

In LED sources, factors that cause lumen depreciation include drive current and heat generated within the device itself (technically speaking, at the diode's p-n junction), which degrades the diode material. Some white-light LEDs may experience degradation of the phosphor coating like that of fluorescent lamps. Some LEDs can also lose lumen output due to clouding of or impurities in the encapsulant used to cover LED chips.

Lumen maintenance measurements take the form Lp, where L is the initial output of a light source, and p is the percentage maintained by the light source over a certain number of hours. L_{97} measures how long a light source retains 97% (or loses 3%) of its initial output, L_{44} measures how long a light source retains 44% (or loses 56%) of its initial output, and so on.

Since high-performance LED light sources can produce useful light for tens of thousands of hours, and since they rarely fail outright, lumen maintenance is often used in place of rated life measurements for LEDs. Measuring the rated life of LED light sources — the mean time to failure of a representative sample — would require operating the sources continually until they finally faded to darkness, a process which would take many years. Because LED light sources continue to deliver light even after their initial lumen output has decreased by 50% or more, lighting specifiers and designers need to know how long an LED lighting fixture will retain a meaningful percentage of its initial light output, not how long it will take for its light sources to fail.

Defining the Useful Life of LED Light Sources

The Alliance for Solid State Illumination Systems and Technologies (ASSIST), a group led by the Lighting Research Center at Rensselaer Polytechnic Institute in Troy, New York, has published a series of recommendations defining the *useful life* of LED light sources. ASSIST defines useful life as the length of time a light source delivers a minimum acceptable level of light in a given application.

> ✴ *LED light sources deliver light even after their initial output has decreased by 50% or more. Lighting specifiers, therefore, need to know how long an LED light source will retain a meaningful percentage of its initial output, not mean time to failure.*

Research performed by ASSIST indicates that changes in general office lighting levels go largely undetected as long as light levels stay above 70% of their initial levels, especially if the changes are gradual. For general lighting applications, therefore, ASSIST recommends defining useful life as the length of time it takes an LED light source to reach 70% of its initial light output (L_{70}). For decorative and accent applications, ASSIST recommends defining useful life as the length of time it takes an LED light source to reach 50% of its initial output (L_{50}).

L_{70} and L_{50} are widely used by the LED lighting community as two important thresholds for useful life, covering a wide range of lighting applications.

The Lumen Maintenance Gap

All well and good — so far. But there's a disconnect between the test results typically provided by LM-80 on the one hand and the L_{70} and L_{50} thresholds that define useful life on the other. This disconnect, which could be called the lumen maintenance gap, is the source of a fair amount of confusion among

> ## Lumen Maintenance of LED and Traditional Light Sources
>
> When properly controlled and driven, LED light sources can have useful lives that last considerably longer than the rated lives of conventional sources. The following table compares the typical useful life range of LED light sources with the typical rated life ranges of conventional light sources.
>
> | Light Source | Typical Range (Hours) |
> | --- | --- |
> | Incandescent | 750 – 2,000 / rated life |
> | Halogen Incandescent | 2,000 – 4,000 / rated life |
> | CFL | 8,000 – 10,000 / rated life |
> | Metal halide | 7,500 – 20,000 / rated life |
> | Linear Fluorescent | 20,000 – 30,000 / rated life |
> | White-light LED | 35,000 – 50,000 / useful life (L_{70}) |
>
> Source: US Department of Energy

lighting specifiers, designers, and other lighting professionals who need to understand how long an LED lighting system will deliver effective light in a particular application. This understanding is crucial for making valid comparisons between conventional and LED lighting fixtures, and for accurately calculating installation, maintenance, and replacement costs. Let's see if we can sort things out.

LM-80 requires testing of LED light sources for 6,000 hours, and recommends testing for 10,000 hours. It calls for testing LED sources at three junction temperatures — 55° C, 85° C, and a third temperature to be determined by the manufacturer — so that users can see the effects of temperature on light output, and it specifies additional test conditions to ensure consistent and comparable results.

Unfortunately, LM-80 provides no recommendations on how to extrapolate measured data to L_{70} or L_{50}. Such a methodology, IES Technical Memorandum TM-21, is currently under development. Until TM-21 is published, the only way an LED source manufacturer can claim that their L_{70} and L_{50} figures conform to LM-80 is to measure their LED sources until they reach those thresholds. Since a typical L_{70} number is 50,000 hours, such a test would last longer than five years! Not only would this test be impractical, but LED technology evolves so quickly that a given product would be obsolete by the time the test was completed.

In practice, leading LED source manufacturers test their products to the LM-80 minimums of 6,000 or 10,000 hours, then apply their own extrapolation methodologies to arrive at L_{70} and L_{50} figures. Since these methodologies are proprietary, manufacturers can choose to disclose as much or as little of the mathematics and supporting data as they wish.

For example, a leading LED source manufacturer publishes the raw data for its high-performance white-light LED in an LM-80 test report. The report includes data on a significant sampling of devices, each tested to 6,000 hours in accordance with LM-80 methods, and L_{70} extrapolations based on an exponential model. While this set of data is sufficient to establish the manufacturer's credibility, users and specifiers would benefit from more transparency into the model's extrapolation formulas and assumptions.

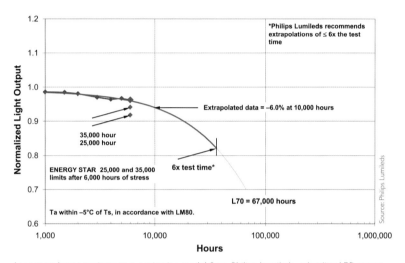

Long-term lumen maintenance projection model from Philips Lumileds, a leading LED source manufacturer, based on LM-80 test results. LED sources are tested for 6,000 hours, and long-term lumen maintenance numbers are extrapolated.

Another leading LED source manufacturer bases the lumen maintenance model for a high-performance white-light LED on their interpretation of raw LM-80 test data. According to their published specifications, the data indicate that lumen maintenance is linear after the first 5,000 hours of operation, so they apply a linear model using variables such as the temperature of the thermal pad on the bottom of the LED, junction temperature, ambient temperature, and drive current. While they do not disclose their extrapolation formulas or raw test data, they do clearly explain their approach, and they provide an extensive set of charts to show expected lumen maintenance to L_{70} at different ambient temperatures and drive currents.

Regardless of the extrapolation method used, keep in mind that L_{70} and L_{50} figures may be *based* on LM-80 measurements, but they are *not* LM-80 measurements.

The Useful Life of LED Sources in Lighting Fixtures

The approved method for making photometric measurements of LED lighting products specifically calls for the testing of complete LED lighting *fixtures* (as spelled out in IES LM-79-08). The approved method for measuring lumen maintenance is just the opposite: It calls for the testing of LED *light sources*, not complete LED lighting fixtures. LM-80 explicitly defines light sources as "packages, arrays, and modules only." This means that LED fixture manufacturers must define their own methods of calculating lumen maintenance for their LED lighting fixtures. As with L_{70} and L_{50} figures provided by LED source manufacturers, lumen maintenance figures provided by LED fixture manufacturers may *use* LM-80 test data and lumen maintenance extrapolations based on them, but they are *not* LM-80 measurements.

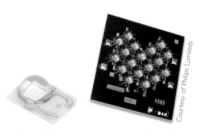

Left, an LED package consisting of a single LED chip, lens, and substrate. Right, an LED array (module), incorporating multiple LED packages. LM-80 applies to packages, modules, and arrays only, not to complete lighting fixtures.

Ambient and internal operating temperatures and drive currents have a significant effect on the lumen maintenance of LED light sources integrated into lighting fixtures, but so do many features of LED lighting fixtures themselves, including lensing, housing color, quality of components, and thermal design. Operational factors such as power surges, static discharge, vibration, and moisture infiltration can also have a significant effect. LM-80 testing for complete LED lighting fixtures would be prohibitively complex and expensive for manufacturers, as they would have to test every different version of their fixtures to account for the effect of each feature or combination of features.

In practice, therefore, reputable LED fixture manufacturers ensure that their fixture drive currents and operating temperatures (especially junction temperatures) fall within the ranges specified by the LED source manufacturers in their lumen maintenance reports. The fixture manufacturers then make their own calculations of the useful life of the LED sources integrated into their lighting fixtures, based on their understanding of the effects of specific physical and operational features.

Useful Life Is Not Fixture Lifetime

It's important to keep in mind that *useful life* and *fixture lifetime* are two very different things. The useful life of a fixture refers to the lumen maintenance projections of the LED sources integrated into that fixture — that is, the number of hours an LED lighting fixture will deliver a sufficient amount of light in a given application.

Fixture lifetime, on the other hand, has to do with the reliability of the components of an LED lighting fixture as a system, including the electronics,

The long useful life of LED sources allow the use of white and color-changing light fixtures in locations where lamp maintenance may be problematic or impossible. Unusual exterior architectural installations, such as the 1,815 ft (553 m) high CN Tower in Toronto, Canada, use LED lighting fixtures to dramatically reduce maintenance labor and costs. The ColorBlast 12 fixtures in use today can natively produce millions of colors, and will maintain 50% of their original lumen output over 50,000 hours or more of use.

Photography: George Fischer

materials, housing, wiring, connectors, seals, and so on. The entire system lasts only as long as the critical component with the shortest life, whether that critical component is a weather seal, an optical element, an LED, or something else. From this point of view, LED light sources are simply one critical component among many — although they are often the most reliable component of the whole lighting system.

Reputable LED lighting fixture manufacturers spend a great deal of time and effort designing and developing all aspects of a lighting system, including control algorithms, board layouts, component quality, thermal management features, optics, and mechanical design. The LED lighting fixture design is then typically validated through a series of in situ tests to verify that the fixture is meeting the expected performance levels for heat dissipation, light output, and so on. Since all the aspects of an LED lighting fixture are interdependent, operational performance can be determined only by testing the fixture as an integrated system.

Comparing the Useful Life of Conventional Lamps and LED Lighting Fixtures

Since all electric light sources experience lumen depreciation, it ought to be possible to extrapolate rated life and lumen depreciation figures for conventional light sources to arrive at L_{70} (and L_{50}) figures. This would allow lighting designers and specifiers to compare apples with apples — that is to say, the useful life of LED sources as incorporated in lighting fixtures with the useful life of incandescent or fluorescent lamps.

- **Incandescent Comparison** A 60-watt incandescent lamp has a rated average life of 1,000 hours. If we assume typical light loss of 10% – 15% over its life, the lamp will fail before reaching the L_{70} threshold. Therefore, its rated life is effectively its useful life.

- **CFL Comparison** An 18-watt CFL lamp has a rated life of 15,000 hours, 1,250 initial lamp lumens, and 1,125 design lumens, which represents a 10% loss after 6,000 hours of operation. The lamp will therefore reach L_{70} after 18,000 hours. But since the lamp is expected to fail after 15,000 hours, its rated life is effectively its useful life.

- **Fluorescent Tube Comparison** High-performance fluorescent lamps are now available with significantly extended rated lives. For example, a 48-inch, 32-watt T8 with an average rated life of 33,000 hours loses 5% of its initial light output after 13,200 hours, or 40% of its rated life. At a constant rate of lumen depreciation, the lamp would reach L_{70} at 79,200 of operation — a figure that rivals the useful life of many LED light sources. Still, the lamp is expected to fail after 33,000 hours, long before it reaches the L_{70} threshold. Therefore, its rated life — not its estimated L_{70} threshold — is effectively its useful life.

- **HID Lamp Comparison** A high pressure sodium lamp, used for streetlighting and outdoor area illumination, lists a rated average life of 24,000 hours. Unlike rated life for the fluorescent and incandescent examples above, rated life for this lamp is based on survival of 67% of the tested lamps (instead of 50%). At 9,600 hours of operation, the lamp maintains 90% of its initial lumen output, which puts its L_{70} threshold at 28,800 hours. While the lamp's rated life and L_{70} measurements are roughly equivalent, the rated average life is still slightly lower. Again, its rated life is effectively its useful life.

As these examples demonstrate, rated life is generally equivalent to useful life for conventional light sources, since conventional light sources typically fail before they reach the relevant lumen maintenance thresholds. Comparing useful life figures for LED lighting fixtures with rated life figures for conventional lamps, then, affords a valid evaluation of how many relampings you can avoid by using LED-based alternatives to conventional lighting solutions. This evaluation in turn offers important information for total cost of ownership comparisons.

For example, eW Cove Powercore, a linear LED cove light from Philips Color Kinetics, reports an L_{70} figure of 60,000 hours. Using eW Cove Powercore instead of the T8 fluorescent fixture in the example above avoids two relampings, four relampings if used instead of the CFL lamp, and 60 relampings if used instead of the incandescent lamp. A Philips Gardco Radiant LED area luminaire, with energy consumption and light output similar to the high pressure sodium streetlamp in the example above, reports L_{70} figures ranging from 50,000 to 100,000 hours, depending on ambient operating temperature and drive current. The LED alternative in this case, therefore, offers a useful life two to four times longer than the HID source.

Getting Dependable, Accurate Information

Given the lack of transparency in the lumen maintenance projections of both LED source and fixture manufacturers, how can lighting specifiers and designers evaluate whether a fixture manufacturer's useful life figures are accurate?

To begin with, specifiers and designers should always look for a reputable fixture manufacturer with a proven track record, and should make sure that the manufacturer offers a comprehensive set of published specifications, photometric data, and related information. But remember that lighting professionals can't simply ask an LED fixture manufacturer for its LM-80 data: As we've seen, only LED *source* manufacturers obtain LM-80 data, and only for a period of operation significantly shorter than accepted useful life thresholds.

Nevertheless, users can increase their confidence in an LED fixture manufacturer's useful life figures by making sure that:

- The manufacturer of the LED sources incorporated in an LED lighting fixture were tested in accordance with LM-80

- The LED source manufacturer uses a valid method of projecting LM-80 test results to L_{70} and L_{50}, based on recommended operating conditions

- The LED lighting fixture manufacturer performs their own measurements of junction temperature, drive current, and other relevant factors, and bases their fixture's L_{70} and L_{50} figures on LM-80 extrapolations provided by the LED source manufacturer

- The LED fixture manufacturer bases its published photometric data on test results from an independent NIST-traceable testing lab, using absolute photometry in accordance with methods and conditions spell out in LM-79

Leading LED fixture manufacturers design their fixtures to ensure that they are as durable and reliable as possible. Although fixture failures do sometimes occur, well-designed LED lighting fixtures can perform reliably for many thousands of hours, often until the LED sources within them have reached the end of their useful life.

Driving and Powering LED Lighting Fixtures

Understanding LED lighting power and control options can help answer such questions as these:

- How easy is an LED lighting fixture to install?

- Is the fixture dimmable, and if so, how?

- Does the fixture require additional power supplies, and if so how many?

- Does the fixture require a separate controller?

- What sort of controllers are available, and how complicated are they to use?

- Can a particular LED lighting fixture be used in retrofit applications?

LED Drivers

In LEDs, current increases very quickly as voltage increases. Small fluctuations in voltage therefore cause large variations in current, which can damage LEDs. In order to connect LED lighting fixtures to a voltage source such as mains power or a battery without damaging the LED sources, input power must be controlled so that the LEDs can safely use it. This is job of the *LED driver*.

An LED driver is an electronic circuit which converts input power into a *current source* — a source in which current remains constant despite fluctuations in voltage. An LED driver protects LEDs from normal voltage fluctuations, as well as from overvoltages and voltage spikes.

LED lighting fixtures that feature integrated LED drivers are as easy to connect to power as any conventional lighting fixture. An increasing number of integrated drivers for white-light LED fixtures are dimmable.

Power Options for LED Lighting Fixtures

Power options for LED lighting fixtures afford performance, ease-of-use, or cost advantages for particular applications. Three common power options are:

- Low-voltage power distribution
- Onboard power integration
- Inboard power integration

Low Voltage Power Distribution

Low-voltage LED fixtures require low-voltage power supplies or transformers and special cabling to convert line voltage into low voltage. A low-voltage power supply is essentially a "brick" in the power cord, similar to a laptop computer's power supply, and it usually produces direct current (DC).

Low-voltage systems are relatively inefficient because power is lost in the conversion from line voltage to low voltage, often through a series of modules for buck conversion, filtering, and processing. Nevertheless, low-voltage systems are preferable for certain types of applications. The rental, touring, and entertainment industries favor low-voltage lighting fixtures, which are often combined with other low-voltage devices for controlling light and sound in theatrical productions. Low-voltage

ColorBlast 12 and its companion fixture, ColorBlast 6, are low-voltage, color-changing LED fixtures from Philips Color Kinetics. ColorBlast 12 is powered by a low-voltage power supply wired to mains.

fixtures can sometimes address aesthetic concerns as well — for instance, in stage designs which require no visible cabling, low-voltage fixtures can be run wirelessly with battery packs and RF controllers.

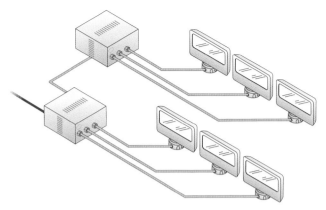

A typical low-voltage configuration consists of one or more power supplies providing power and control to one or more lighting fixtures. Here, PDS-150e power supplies run three ColorBlast 12 fixtures each.

A typical low-voltage configuration is a "star" configuration, where each fixture or run of fixtures connects directly to a low-voltage power supply through a unified power cable, often a proprietary leader cable designed to work with a specific fixture. The power supply, in turn, is connected to a power source. The number of fixtures that can be attached to each power supply is limited by such factors as the fixtures' power consumption, the distance between the fixtures and the power supply, and the number of available power supply ports.

ColorBlast 12 is a color-changing, low-voltage LED floodlight from Philips Color Kinetics, often used for wall-washing in stage lighting situations. Up to three fixtures can be connected to a single power supply, each on a maximum cable run of 60 feet. Installations requiring many fixtures use multiple power supplies, each positioned appropriately in relation to the fixtures, and each connected to a power source.

ColorBlaze®, a powerful, full-color linear LED fixture for theatrical and stage applications from Philips Color Kinetics, features onboard power integration, thermal sensors, and cooling fans. ColorBlaze connects to power with a standard IEC power cable.

Onboard Power Integration

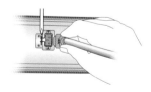

Onboard power integration uses the same overall control scheme of traditional low-power systems but offers a few advantages. It replaces external low-voltage power supplies with standard switching power supplies integrated directly into the fixtures, allowing the fixtures to be connected directly to line voltage. This approach can save on setup and installation costs, but the additional fixture components can increase fixture size and thermal load.

ColorBlaze®, from Philips Color Kinetics, is a high-output linear LED fixture for theatrical use that features onboard power supplies and related thermal management controls, such as onboard thermal sensors and cooling fans. Color-Blaze can be installed simply by connecting it to line voltage with a standard IEC power cable.

Inboard Power Integration

Inboard power integration represents an entirely different approach to power management. Inboard power integration schemes incorporate the power supply directly into the fixture's circuitry to create an efficient power stage that consolidates line voltage conversion and LED current regulation. By integrating a single, efficient power stage into the LED lighting fixture itself, inboard power integration can eliminate a significant percentage of the power losses associated with low-voltage configurations with multiple power stages.

Where inboard power integration is appropriate, users can reap many advantages, including increased system efficiency and lower cost and complexity of installation, operation, and maintenance.

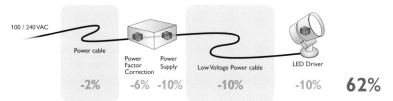

Low-Voltage Power Distribution

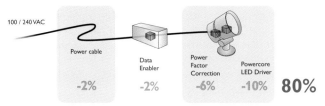

Inboard Power Integration for Color-Changing Fixture (Powercore)

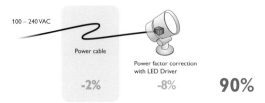

Inboard Power Integration for White-Light Fixture (Powercore)

By integrating the power supply and power factor correction modules with the LED drivers directly in the lighting fixture's circuitry, Powercore eliminates some of the losses associated with low-voltage power distribution systems.

Powercore, a microprocessor-based inboard power integration system patented by Philips Color Kinetics, is used in many of its line-voltage LED lighting fixtures. Powercore technology is state-of-the-art for fully integrated power management of LED lighting fixtures:

- Powercore can eliminate from 18% to 30% of the power losses associated with external low-voltage power supplies and cabling.

- Powercore integrates active power factor correction (PFC) circuitry into each fixture to lower operating costs and maximize operational efficiency. Power factor in Powercore-based lighting fixtures typically measures above 0.995, very close to the ideal power factor of 1.000.

- Powercore enables universal power input. Powercore fixtures can receive input voltages in the range of 100 – 240 VAC, and reliably and efficiently supply the required wattage to run the fixture.

- By minimizing the number of required power supplies, Powercore simplifies and lowers the cost of installation and maintenance.

Controlling LED Lighting Fixtures

Control is a general term for a wide range of methods, protocols, and devices for operating LED lighting fixtures. The simplest forms of control are ON / OFF switching and dimming. For many single-color and static white light LED fixtures, these are the only applicable methods of control.

Full-color and tunable white light LED fixtures can be controlled to display any available RGB color or color temperature, simple color-changing patterns, intricate light shows authored by end-users or professional show designers, and even large-scale video. Dynamic LED lighting fixtures typically accept input from specially designed controllers using a *communications protocol* that the lights can understand. A communications protocol is simply a standard set of rules for sending signals and information over a communications channel.

DMX Control

As the lighting industry has evolved, so has the range of available communications protocols. The most commonly used and accepted control format for color-changing lights is DMX512-A, or DMX for short. DMX was originally developed by the Engineering Commission of the United States Institute for Theatre Technology (USITT), beginning in 1986, for controlling stage and theatrical lighting. Most theatrical lighting boards are DMX-based, but they tend to be too large, complex, specialized, and expensive for general use. Some manufacturers of dynamic LED lighting fixtures for non-theatrical use, therefore, develop and market their own DMX-based controllers. These are generally more compact and less expensive than full-fledged theatrical lighting boards, and they often offer special features — such as pre-defined light shows and built-in effects — designed to simplify and automate lighting control for general users.

Very briefly, DMX-based controllers communicate with LED lighting fixtures using DMX *addresses*. Each lighting fixture in an installation is assigned an address, or a set of addresses. These addresses allow a controller to identify individual lighting fixtures within the installation and send the fixtures specific control signals so that each fixture can display the correct light output.

Uniquely addressing and controlling color-changing LED lighting fixtures lets you display different light output — different colors and different brightnesses — on multiple fixtures simultaneously. This level of control enables an infinite variety and combination of dynamic effects, from colors that fade one into another or that seem to chase each other from fixture to fixture, to intricate light shows that mimic the appearance of natural phenomena or that display abstract patterns for subtle or dramatic effect.

Most color-changing LED fixtures have three channels, one for each color of LED used in the fixture — usually red, green, and blue. Each fixture receives three separate channels of DMX data from the controller, one for the red LEDs, one for the green, and one for the blue. The first fixture in an installation could be programmed to receive DMX data via addresses 1, 2, and 3; the second could be programmed to receive DMX data via addresses 4, 5, and 6; and so on. DMX supports a maximum of 512 DMX addresses. A single DMX *universe* consists of a maximum of 170 uniquely addressed three-channel fixtures (512 divided by 3 = 170, with two channels left over). An installation can consist of one or more DMX universes.

iPlayer® 3, from Philips Color Kinetics, is a compact LED lighting controller capable of controlling two separate DMX universes of 512 addresses each.

iPlayer® 3 from Philips Color Kinetics is a compact but powerful DMX controller for full-color LED lighting fixtures. Like many DMX-based controllers, iPlayer allows you to store multiple light shows on its internal memory (exactly as you would store files on a computer). You can select, start, and stop the stored light shows and otherwise control connected lighting fixtures with the push-button controls built into the iPlayer 3, or you can use a remote keypad designed to work with iPlayer 3. Along with 10 customizable effects — including chasing, ripple, burst, and cross-fade effects — iPlayer also comes with Color-Play® 3 show-designing software, which runs on any Macintosh or Windows computer.

iColor Flex SLX, a string of 50 individually controllable full-color LED nodes from Philips Color Kinetics, can be used to display color-changing effects and large-scale video on two- and three-dimensional surfaces.

Ethernet Control

Since LED lighting fixtures are inherently digital, entire lighting systems can be run on Ethernet computer networks. Ethernet-based systems do not have the same addressing limitations as DMX, and so are preferable for larger installations. Ethernet control is effectively required for large-scale video installations, where potentially thousands or tens of thousands of LED fixtures must be individually addressed and controlled.

Direct-view LED lighting fixtures designed specifically for displaying video sometimes contain multiple individually controllable segments, or *nodes*. iColor Flex SLX from Philips Color Kinetics, for instance, is a strand of 50 individually controllable full-color LED nodes. Large-scale video installations can be constructed from multiple iColor Flex SLX strands mounted on a two-dimensional or three-dimensional surface.

The Stadion Center, a landmark shopping destination in Vienna, Austria, uses 37,620 individually addressed iColor Flex SLX nodes to display advertisements, video, and dynamic full-color effects across its curved façade each night. Some Ethernet controllers, such as Video System Manager Pro from Philips Color Kinetics, can address and control up to 250,000 unique LED nodes, each with three channels.

Photo courtesy of Matthias Silveri

Other Control Options

DMX512 and Ethernet are two of the most widely used communications protocols for controlling LED lighting systems, but there are others:

- In Europe, the Digital Addressable Lighting Interface (DALI) is a commonly used alternative to DMX.

- Some companies have developed their own proprietary Ethernet-based communications protocols, such as KiNET for Philips Color Kinetics lighting fixtures.

- ACN and Streaming ACN are the Entertainment Services & Technology Association (ESTA) standard DMX over Ethernet protocols. Streaming ACN is a scalable method of transmitting multiple universes of DMX data.

These are only a few of the available choices. As often happens with a proliferation of options, compatibility is a concern. Conversion devices are sometimes available to connect lighting fixtures that require one communications protocol with controllers that require a different protocol. The situation can get complicated quickly, so it's often simplest to use components that are designed to work together.

Dimming LED Lighting Fixtures

LED lighting fixtures can be dimmed in two different ways, depending on the type and capabilities of the fixtures:

- Color-changing and tunable white LED lighting fixtures can be dimmed via a DMX or other control interface.

- Solid-color and solid white light fixtures can be dimmed via a compatible, commercially available dimmer.

Dimming LED Lighting Fixtures Via DMX or Another Control Interface

As described above, color-changing and tunable white light LED fixtures usually accept DMX control or another communications protocol designed for controlling lighting fixtures. Any of these control interfaces can dim LED fixtures by reducing the output equally across all channels.

The growing number of white-light LED products with dimmable integrated LED drivers includes linear cove lights, such as eW Cove MX Powercore from Philips Color Kinetics, and incandescent replacement lamps such as the Philips MASTER LEDbulb A55.

In practice, dimming controls are usually available as standard on controllers, keypads, and other devices designed specifically for use with certain LED lighting fixtures. iW® Scene Controller and ColorDial™ Pro, two simple wall-mounted controllers from Philips Color Kinetics, offer push-button dimming controls for tunable white and full-color LED lighting fixtures, respectively.

Dimming LED Lighting Fixtures with Commercially Available Dimmers

Solid color and solid white light LED lighting fixtures with integrated dimmable drivers can be dimmed with compatible, commercially available dimmers. Most LED drivers use pulse width modulation (PWM) to regulate the amount of power to the LEDs. Like incandescent lamp dimmers, PWM turns LEDs on and off at high frequency, reducing total ON time to achieve the desired dimming level.

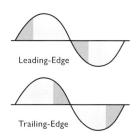

Leading-edge dimmers chop the beginning of each half-cycle of AC power, while trailing-edge dimmers chop the end of each half-cycle.

Typical incandescent lamp dimmers work like switches that toggle on and off 120 times per second. By "chopping" the beginning of each AC power waveform, they regulate the amount of power to the lamp. The waveform chopping happens very rapidly, so most people see continuous dimming without flicker. This sort of dimming scheme is known as *leading-edge* dimming.

Most LED drivers are incompatible with incandescent lamp dimmers, and with leading-edge dimmers in general. LED fixtures con-

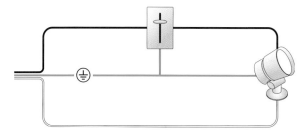

Two-Wire Dimmer Circuit

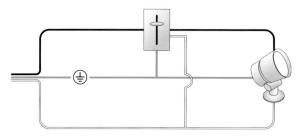

Three-Wire Dimmer Circuit

nected directly to incandescent dimmers can visibly flicker at low dimming levels. Worse, voltage spikes can damage the fixtures.

Many LED lighting fixtures that work with switching power supplies work best with electronic low voltage (ELV-type) dimmers. Fixtures that work with magnetic transformers, such as MR16-compatible LED lamps, typically require magnetic low voltage (MLV-type) dimmers.

Most ELV-type dimmers are *trailing-edge* — that is, they chop the end of each AC power waveform. In general, trailing-edge dimmers perform much more reliably with LED lighting fixtures. ELV dimmers are wired similarly to standard dimmers, except they have an "extra neutral" wire that maintains power to the dimmer even if the lights connected to it are turned off.

Trailing-edge dimmers are in much less common use than leading-edge dimmers. In new installations, this poses no problem — you simply install the recommended dimmer type along with the new lighting fixtures. In retrofit installations, however, where you are replacing conventional lighting fixtures with LED lighting fixtures, you may have to swap existing leading-edge dimmers for compatible trailing-edge dimmers.

Choosing the right dimmer is important to avoid flickering and ghosting, which occurs when fixtures produce visible light even though the lights are turned off. It's a fairly safe bet that a particular LED fixture will work well with only a few compatible dimmers. The fixture's driver design determines which commercially-available dimmers will work best. LED lighting fixture manufacturers usually publish lists of tested and approved dimmers, and only these should be used.

Dimming Threshold and Dimmer Wattage

The effective dimming threshold of most LED lighting fixtures is around 10% and varies with the particular dimmer being used. Most commercially available dimmers have a trim-pot screw that controls the minimum dimming level. The wattage of the dimmer you select must be sufficient for your lighting installation. To determine minimum dimmer wattage, you can multiply the number of fixtures the dimmer needs to control by the wattage of each fixture. If your installation uses a mix of fixtures, you can multiply the wattage of each type separately, then add the results to get the minimum dimmer wattage.

High-resolution 12-bit or 16-bit LED drivers increase the number of dimming "steps" by orders of magnitude, virtually eliminating visible flicker. High-resolution LED fixtures can usually be dimmed below the 10% limit imposed by most 8-bit LED drivers.

What to Look for in LED Lighting Solutions and Manufacturers

With a general but fairly thorough grounding in LED lighting technology, users can make informed, accurate judgments concerning the quality of specific LED lighting fixtures, the reputation of the fixture manufacturer, and the suitability of an LED lighting fixture for a particular application. Users should consider the following features and ask the following questions when selecting LED lighting solutions.

Reputation Look for a manufacturer with a solid reputation and proven track record in the LED lighting industry.

Support Make sure that the LED lighting manufacturer offers online and phone-based technical and application support, and that it has a return / repair program in the case of fixture damage or failure.

Systems Thinking The most effective LED lighting fixtures and solutions result from systems thinking — conceiving of lighting fixtures, accessories, and power and control options as a complete system in which each component functions as part of an integrated whole.

Portfolio of Products You have the best chance of success when selecting a manufacturer with a complete portfolio of products.

Control Solutions Look for manufacturers that provide integrated control solutions, specifically designed and optimized to work with the LED lighting fixtures they produce.

Thermal Management The mechanical fixture design should properly address heat dissipation, and the fixture should feature firmware linked to junction temperature feedback. The best LED manufacturers require military spec / grade for many key components.

Dimming Make sure white-light LED fixtures are smoothly and reliably dimmable with standard, commercially available dimmers.

Standards Testing LED lighting manufacturers should use an independent, NIST-traceable testing lab, publish accurate, complete specifications and photometric data, and make test lab data files and reports available for evaluation.

Quality of Light LED lighting fixtures should conform to or exceed ANSI binning standards for consistency in flux, color or color temperature, and forward voltage. White-light LEDs should render colors accurately and have high efficacy.

Useful Life Useful life should be calculated and reported in fixture data sheets. Well-designed fixtures target junction temperatures that strike the optimum balance between light output and longevity.

Weatherability Weatherability can range widely depending on fixture type, quality of fixture housing design, and consistency of manufacturing processes. IP ratings are important, but do not guarantee long-term durability.

Electrical Design Well-designed fixtures have robust electrical design. Look for features such as short circuit protection, surge protection, and reverse polarity protection.

4

LED Lighting Applications

LED LIGHTING FIXTURES are available for virtually any lighting application. Often, LED lighting alternatives offer advantages over comparable conventional fixtures — including superior energy efficiency, useful lifetimes up to 50 times longer, low maintenance and replacement costs, and a range of color temperatures or millions of colors from a single fixture.

> ✹ Because of their flexibility and range of features, many LED fixtures can support applications in multiple areas. Look to manufacturers' specification sheets, product guides, brochures, and other publications for more detailed information on specific fixtures and applications.

This chapter looks at 10 major application areas, profiles one or more LED fixtures especially suitable for each area, and presents case studies that demonstrate LED lighting fixtures in successful applications around the world today.

Task Lighting

In offices and other workspaces, task lighting is intended to deliver the proper quality and quantity of light to a task area. The task area can be a surface for performing tasks requiring concentrated illumination, high contrast, or accurate color rendering. Typical tasks include reading and writing at a desk, preparing food at a kitchen counter, graphic design at a drafting table, or working with electronics or small machinery on a workbench. *The IESNA Lighting Handbook* recommends from 30 – 50 fc (300 – 500 lx) for performing most tasks involving high contrast, and up to 100 fc (1000 lx) for tasks that are both low in contrast and small in size.[33]

Some workspace lighting schemes use downlighting or other ambient lighting to illuminate task areas, some use adjustable attached or free-standing lamps, and some use runs of linear under-cabinet fixtures. LED lighting fixtures excel in this last area. Fixed or adjustable color temperatures can ensure a sufficient amount of contrast. Integrated primary optics afford tightly controlled beam angles that deliver illumination to the task area with very little wasted light and virtually no glare.

eW Profile Powercore

eW Profile Powercore, from Philips Color Kinetics, is a direct line voltage under-cabinet LED fixture for common task lighting applications. The warm 2700 K option is appropriate for incandescent and fluorescent retrofits or replacements, and the cool 4000 K option can be used in place of fluorescents and metal halide fixtures.

eW Profile Powercore, from Philips Color Kinetics, is an energy-efficient, under-cabinet LED task light appropriate for most common applications requiring high contrast and accurate color rendition.

Consuming only about 6.2 watts per foot, eW Profile Powercore uses 80% less energy than a comparable xenon under-cabinet fixture, and 40% less energy than a comparable T8 fluorescent fixture. Direct line voltage input with integrated inboard power eliminates off-state power consumption, further increasing efficacy.

eW Profile Powercore can deliver 40 – 50 fc (400 – 500 lx) in typical under-cabinet configurations, making the fixture appropriate for most high-contrast tasks. Its integrated 110° x 110° lens is optimized for delivering light to task areas, following IESNA recommendations. A 20° beam offset is built into the housing design so that fixtures can be installed flush to mounting surfaces without any shimming or fine adjustments. An ultra-low profile housing, and multiple fixture lengths, jumper cables, and connector options can accommodate virtually any layout.

Case Study: Under-Cabinet Lighting / Private Residence

The advantages of innovative, energy-efficient white-light LED sources are on display in a high-end kitchen in a private residence in Cambridge, Massachusetts. The homeowners wanted to create a hip, modern lighted environment without sacrificing energy efficiency.

Fluorescent fixtures were replaced with eW Profile Powercore fixtures to illuminate the kitchen's generous counter space. The lighting designer installed eight 19.25 in, 4000 K fixtures underneath the upper cabinets. The fixtures' high CRI of 80 is an important factor for food preparation. With their ultra-thin housing (only .875 in thick and 1.5 in wide), the fixtures are virtually invisible, creating a clean, modern look and eliminating direct glare.

Eight 400 K eW Profile Powercore under-cabinet LED fixtures were installed in this private kitchen in Cambridge, Massachusetts. Fixtures are 40% more energy efficient than the T8 fluorescents they replaced, and deliver 40 – 50 fc (400 – 500 lx) and CRI of 80 in this application, for sufficient quality and quantity of light.

Downlighting

As a general term in lighting, *downlighting* refers to a number of techniques for providing ambient white lighting. In some ways, downlighting is the opposite of task lighting: Its role is to provide overall lighting in a room or other interior space, rather than highly directed light targeted at a specific task area.

Usually, the goal of ambient lighting is to provide enough light for basic visual recognition and for moving around in a space. The degree of contrast between ambient light levels and task, accent, and other light levels can have a significant effect on the atmosphere and emotional impact of a space. A high degree of contrast can make a space feel dramatic, while a low degree of contrast can make a space feel cheerful and relaxing. Downlighting that illuminates the floor and not the walls or ceilings can increase the sense of drama.

The term *downlight* refers specifically to downlighting fixtures installed in an opening in or mounted flush to a ceiling. Downlights are usually either round or square, and have different apertures and beam angles for delivering a broad spread or narrowly focused column of light. As we saw in Chapter 3, well-designed LED downlights can offer light output equivalent to or better than conventional incandescent and CFL downlights, with significantly better energy efficiency. LED downlights typically have form factors and installation requirements similar to their conventional counterparts, so they are easy to spec and install in new and retrofit applications.

eW Downlight Powercore

eW Downlight Powercore, from Philips Color Kinetics, is a dimmable, direct-line voltage, surface-mount LED downlight suitable for illuminating corridors, lobbies, and other common areas in an office space. Installation is made simple, with power modules in a range of input voltages, LED modules with a selection of beam angles and square bezels with a variety of finishes for matching many interior decors. Warm 2700 K fixtures are appropriate for intimate, open environments such as restaurants, hotel lobbies, and homes, and cool 4000 K fixtures for lighting clean and efficient spaces such as offices, classrooms, and hospitals.

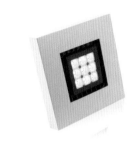

eW Downlight Powercore, from Philips Color Kinetics, is a dimmable, direct-line voltage, surface-mounted LED downlight with light output comparable to CFL downlights and superior lifetime and energy efficiency.

As we saw in Chapter 3, eW Downlight Powercore delivers light levels and quality of light comparable to CFL downlights with no wasted energy, light, or heat. As an ENERGY STAR qualified LED luminaire, eW Downlight Powercore uses 80% less energy and can last over 40 times longer than incandescent lighting — up to 85,000 hours of use at 70% lumen maintenance.

Calculite LED Downlight

Calculite LED Downlight, from Philips Lightolier, is an award-winning LED downlight that uses a *remote phosphor* system for creating white light. White-light LEDs typically use a phosphor-coated blue LED chip. Because the blue light is scattered on contact with the phosphor, up to 60% of the converted light is back-reflected. Calculite remote-mounts the phosphor to a diffusion lens, allowing

Calculite LED Downlight, from Philips Lightolier, uses a remote-phosphor system to produce consistent, uniform white light at an efficacy of 50 lm / W.

back-reflected light to be redirected out of the luminaire via a highly reflective mixing chamber. The result is a 20% increase in system efficiency, for an industry-leading efficacy of 50 lm / W.

Calculite fixtures are modular. The proprietary power supply, which rapidly and accurately controls power to the fixture's LEDs, is separated from the light engine to minimize heat conduction. Both power supply and light engine can be independently and easily accessed below-ceiling. Calculite downlights can be dimmed with selected off-the-shelf dimmers, to as low as 5% of total output depending on the dimmer.

New construction frame-in kits offer multiple finishes, apertures, color temperatures, and input voltages. Retrofit frame-in kits allow simple conversion of existing Calculite incandescent, CFL, and HID downlights.

Case Study: Downlighting / Retail Space

Heineken added a new dimension to its company by opening a unique, ultra-modern concept store, Heineken The City, in the brewer's home city of Amsterdam. The store sells special products and services — including music, fashion, travel, events, and its signature beer — in six renovated historical buildings.

Heineken The City's revolutionary, hypermodern design makes effective use of the latest technology, including speaking mirrors, 3D television screens, an ice wall, and interactive pillars. The store is the first in Europe to be entirely illuminated by LED lighting. The lighting designers found LED lighting to be the perfect choice for general, accent, and decorative applications throughout the store.

General lighting is provided by dozens of surface-mounted eW Downlight Powercore fixtures. Cool 4000 K fixtures illuminate the two-story entrance, while warm 2700 K fixtures illuminate the sound studio for a cozier, more intimate atmosphere. Fixtures with a narrow 30° beam angle are used in areas with high ceilings, while the wide 65° beam angle is used in areas with lower ceilings.

Visitors are pleased and impressed with the Heineken The City's sophisticated but relaxed ambience. The store's owners and managers further benefit from the long lifetime and low energy consumption of the LED fixtures used throughout the store.

Clusters of eW Downlight Powercore, LED downlights from Philips Color Kinetics, and indirect lighting illuminate the entryway at Heineken The City in Amsterdam. Recommended light levels for retail spaces such as these is about 30 fc (300 lx).

Cove Lighting

Cove lighting is a form of indirect lighting where linear fixtures are mounted in a cove. A cove is a ledge, recess, coffer, or similar architectural feature along the sides of a room near ceiling height. White-light cove fixtures are used primarily for *uplighting* — indirect lighting that illuminates the ceiling or the upper portions of the walls (a technique also known as *slot lighting*). Uplighting can create a cheerful and relaxed ambience, especially if there is a low level of contrast between the ambient light levels provided by the cove lights and the light levels provided by task or spotlighting. Solid-color or color-changing cove lights are used primarily for accents and aesthetic effects.

LED-based cove lighting fixtures are particularly well suited for both uplighting and effect lighting. Like all LED lighting fixtures, LED cove lights are directional, so they waste far less light than incandescent or fluorescent cove lights, the lamps of which emit light in all directions. Well-designed LED cove lights typically offer a range of beam angles, with built-in aiming and locking features, for precise control over the spread of light beyond the lip of the cove, and for controlling or eliminating socket shadowing and backsplashing in the cove itself. LED cove lights can offer highly uniform fixture-to-fixture mixing, minimizing joins or scalloping between fixtures.

Because LED lighting fixtures can reliably deliver useful light for many thousands of hours, outages that can disrupt linear runs in high-profile locations, such as lobbies and corridors of luxury hotels, are rare. By producing a cool beam of light with no radiated IR or UV, LED cove lighting fixtures can be used in museums, displays, and historical settings where conventional lighting might discolor inks and dyes, or damage fabrics, paints, and other sensitive materials.

Full-color and tunable white LED cove lights offer the ability to create adjustable washes of white or colored light, as well as dynamic, color-changing effects to heighten drama and interest in nightclubs, restaurants, retail spaces, and other locations.

eW Cove QLX Powercore

eW Cove QLX Powercore, a linear, solid white light cove fixture from Philips Color Kinetics, represents an attractive alternative to traditional cove lighting sources. With light output of approximately 300 lumens per foot, a useful life of over 50,000 hours, superior energy efficiency, and virtually maintenance-free operation, eW Cove QLX Powercore can replace comparable T8 and T12 fluorescent sources with a three-year payback, and comparable halogen and xenon sources with one-year payback.

eW Cove QLX Powercore exceeds the ANSI standards for chromaticity, guaranteeing color uniformity

eW Cove QLX Powercore, from Philips Color Kinetics, replaces T8 or T12 fluorescent cove lighting sources with a three-year payback, and comparable incandescent sources with a one-year payback.

and consistency of hue and color temperature within a four-step MacAdam ellipse, allowing variations that are virtually undetectable across LEDs, fixtures, and manufacturing runs.

Like other well-designed LED cove lighting fixtures, eW Cove QLX Powercore offers many features that make mounting, positioning, connecting, and aiming fixtures easy. End-to-end locking power connectors can make 180° turns, fixtures can rotate in 10° increments through a full 180° for precise aiming and color mixing, mounting tracks support vertical and overhead positioning, and jumper cables of various lengths can add extra space between fixtures where needed.

Power connection options include direct, permanent installation to a power source, or portable installations using a standard two-prong plug. eW Cove QLX can be switched, or dimmed with compatible trailing-edge ELV-type dimmers.

iW Cove Powercore

iW® Cove Powercore, from Philips Color Kinetics, is a tunable, intelligent white light cove fixture, offering color temperatures in the 3000 K to 6500 K range. Color temperature can be adjusted independent of brightness, allowing you to select the exact shade and saturation of white light appropriate for any retail, exhibit, hospitality, or architectural application. Like eW Cove QLX Powercore, iW Cove Powercore offers operational efficiency, simple installation, and two beam angles, rotating housing, and flexible end-to-end locking power connectors.

iW Cove Powercore is a tunable white light cove fixture with color temperatures ranging from 3000 K to 6500 K.

iColor Cove MX Powercore

iColor Cove® MX Powercore elevates cove lighting to an entirely new level, offering 16.7 million native colors for creating dynamic, color-changing accents and effects that simply can't be achieved with conventional cove lights. iColor Cove MX Powercore shows how game-changing LED lighting technology can be. In fact, these fixtures can be used for backlighting, inexpensive wall grazing, and direct-view applications, in addition to standard cove lighting applications.

iColor MX Cove Powercore delivers 16.7 million native colors, affording dynamic, color-changing effects that conventional cove lights can't achieve.

Case Study: Cove Lighting / Historical Landmark

Boston's Old North Church has been famous for centuries, ever since Paul Revere's famous midnight ride during the American Revolution. Now its 18th century architecture is illuminated with cutting-edge LED cove lighting. The Old North Church is a compelling example of how even America's oldest buildings can easily and successfully integrate LED lighting.

As part of ongoing renovations, management for the 285-year-old church sought a sustainable and low-maintenance light source to replace the dated, linear incandescent tube system lining 18 interior niches. Accordingly, approximately 130 eW Cove Powercore fixtures were installed to replace the former system. With projected useful life of 50,000 hours, these LED cove lights can last 25 times longer than the incandescent fixtures they replaced, drastically

LED cove lights, installed in interior niches of the Old North Church in Boston, output appealing, warm, and uniform white light, while saving drastically on operational, maintenance, and replacement costs.

reducing replacement and maintenance costs. The new system is expected to cut energy consumption by nearly 85% (40 watts per 8-foot run of the LED system, versus 240 watts per 8-foot run of the incandescent system).

Installed along interior niches of the church's upper gallery, the eW Cove Powercore fixtures cast a warm white glow that accentuates the historic arches and moldings. They can be dimmed using the church's existing ELV-type dimmers. Their low profile and simple line-power installation allows them to fit within the narrow alcoves where light sources that require ballasts, transformers and other auxiliary equipment could not.

Case Study: Cove Lighting / Hospitality Space

Rustic Kitchen Bistro & Lounge emits an inviting glow, largely due to its famous wood burning oven and dramatic lighting.

The restaurant owners wanted warm, attractive lighting to properly accentuate the alcoves within the Biltmore-inspired ceiling in the restaurant's Tuscan Room. They explored various light sources, including neon and fluorescent, but those sources did not offer the desired color temperatures or the long lifetimes and low maintenance benefits of LEDs. Ultimately, the owners chose iW Cove Powercore fixtures for the installation. Since these fixtures provide controllable white light in a wide color temperatures range, the restaurant staff can vary the atmosphere in the dining rooms with the touch of a button – for example, they can dial in a cooler color temperature during the day, and a warmer color temperature in the evening. Unlike conventional light sources, iW Cove Powercore fixtures maintain their precise color temperature even as they are dimmed.

iW Cove Powercore fixtures allow the staff of Boston's Rustic Kitchen to dial in the exact shade of white light they desire — a cooler color for daytime (above), warmer for night (below).

Wall Washing

Wall washing is a technique for evenly illuminating a wall or large surface. Wall washing works best with wallboard or other lightly textured surfaces, as it hides imperfections and tends to flatten the appearance of the illuminated area. (Wall grazing is a technique for bringing out the texture in rough wall surfaces, such as brick and stone — see below.) Wall washing, therefore, is especially common in clean, minimalist interiors, such as modern art museums, new office spaces, and high-end contemporary residences. Wall washing also works best when walls or surfaces are matte, as wall washing of shiny surfaces can create a fair amount of indirect glare.

Like cove lighting and slot lighting, wall washing can make rooms appear brighter, cheerier, and more relaxed. Typical wall washing fixtures have asymmetric lenses, and are positioned some distance from the wall or surface at an angle of incidence intended to minimize glare and shadowing from frames or other objects that may be mounted on the wall or surface.

ColorBlast Powercore

ColorBlast Powercore, from Philips Color Kinetics, is a versatile, full-color LED fixture suitable for interior and exterior wall washing, as well as backlighting, signage, grazing, spotlighting, and floodlighting applications. A wide beam angle throws a soft-edge beam, while narrow beam angles afford for extended light projection for washing high walls and architectural façades.

Like many LED lighting fixtures, ColorBlast Powercore is versatile, supporting a range of applications including wall washing, wall grazing, floodlighting, and spotlighting.

ColorBlaze

ColorBlaze, from Philips Color Kinetics, is a high-performance linear LED fixture for washing large areas with far-reaching, rich, saturated colors in both theatrical and architectural applications. On-board power and addressing makes this fixture especially attractive for rental, touring, and other temporary applications, as its rugged housing and simple IEC power connection support repeated setup, reconfiguration, and teardown. The durability, low maintenance, and intelligent control afforded by LED technology also makes ColorBlaze suitable for permanent architectural installations.

On-board power, fine control, IEC power connection, and rugged housing make ColorBlaze, from Philips Color Kinetics, ideal for rental, theatrical, and architectural wash applications.

Case Study: Wall Washing / Contemporary Landmark

During the 2008 holiday season, the city of Long Beach, California unveiled a series of LED lighting installations illuminating prominent buildings throughout the downtown area. Each landmark was lit with different types of LED fixtures to give each landmark a unique presence and personality.

The Long Beach Convention Center, which was already undergoing major restorations, now uses 100 ColorBlast Powercore fixtures to wash the interior surfaces of its glass atrium with intense, uniform, color-changing light. These fixtures support a number of the design team's goals: They are highly energy-efficient, require little to no maintenance, and natively produce dynamic, colorful effects without the need for colored gels or filters.

Photography: Frank Dorosi

Wall Grazing

Wall grazing is related to wall washing, as it is also a technique for illuminating walls or large surfaces. Unlike wall washing, however, wall grazing is intended to reveal the texture of brick, stone, or other rough or sculpted surfaces. Wall grazing fixtures are typically mounted very close to the surface to be illuminated, in trough or coffer at the surface's base or top. Effective wall grazing fixtures deliver high-intensity light in a relatively narrow beam angle.

Full-color LED wall grazing fixtures offer the ability to graze surfaces with dramatic, dynamic color-changing effects, or to adjust the ambience of a room on the fly for different purposes, seasons, times of day, and so on.

ColorGraze Powercore

ColorGraze Powercore goes far beyond conventional surface grazing light fixtures, offering control resolution down to 1 ft (305 mm) for dynamic effects and light shows.

ColorGraze™ Powercore, from Philips Color Kinetics, is a linear RGB fixture optimized for surface grazing, wallwashing, and signage illumination. ColorGraze Powercore offers uniform beam saturation as close as 6 in (152 mm), and a compact, low-profile design combined with flexible mounting options allows for discreet placement in a wide range of architectural installations. Multiple fixture lengths and beam angles, along with 1 ft (305 mm) light addressing segments within each fixture, allow fine control of color-changing effects and light shows.

Case Study: Wall Grazing / Private Residence

North Dumpling Island, off the Connecticut coast, is a three-acre island owned by prolific inventor Dean Kamen. When the U.S. Coast Guard cut the electrical connections to the island's lighthouse, Kamen took the opportunity to use renewable energy exclusively on the island, together with the latest innovations in lighting, water purification, and appliances — many of which are Kamen's own inventions.

Kamen installed Philips LED lighting fixtures throughout the entire island. Kamen selected ColorGraze Powercore fixtures to illuminate the brick exterior of island's main residence. Eight 4 ft (1219 mm) fixtures were installed end-to-end, discreetly wrapping around the underside of the building's roof overhang. Delivering over 270 lumens per foot, ColorGraze Powercore fixtures drench the textured exterior with washes of vibrant light. The fixtures' superior beam quality allows for striation-free saturation, eliminating visible light scalloping between fixtures.

To support the island's energy efficiency, each ColorGraze Powercore fixture consumes only 70 watts. Unlike the island's former conventional floodlighting system, ColorGraze Powercore fixtures are inherently directional, projecting light only where it's needed, minimizing wasted light and maximizing the system's efficacy.

ColorGraze Powercore graze the brick exterior of the main residence on North Dumpling Island, which uses renewable energy sources exclusively. ColorGraze Powercore supports the island's sustainability with highly efficient, precisely directed light.

Floodlighting

You can think of floodlighting on the one hand as a form of downlighting, and on the other hand as a form of wall washing. High-intensity metal halide fixtures are often used as downlighting for playing fields and other sports venues, or for illuminating large exterior architectural surfaces. Large, ultra-high-intensity LED lighting fixtures are now available that rival the performance and efficacy of conventional high-intensity floodlighting fixtures, especially for colored light.

ColorReach Powercore

ColorReach™ Powercore, a high-performance exterior architectural floodlight from Philips Color Kinetics, is the first LED fixture powerful enough to brilliantly and dynamically illuminate large architectural façades. ColorReach Powercore is specifically designed for large-scale installations, such as commercial skyscrapers, casinos, large retail exteriors, bridges, piers, public monuments, and themed attractions. With and output of over 5,000 lumens and light projection of over 500 ft (152.4 m), ColorReach Powercore represents a true alternative to conventional high-intensity fixtures for exterior illumination. The ability of LEDs to natively produce millions of colors without filtering makes ColorReach Powercore an order of magnitude more efficient in applications calling for colored light.

With output of over 5,000 lumens, light projection of over 500 feet, and the ability to natively produce millions of colors without gels or filters, ColorReach Powercore is the first true energy-efficient alternative to conventional high-intensity floodlighting fixtures.

Case Study: Exterior Architectural Floodlighting / Public Building

Once the headquarters for local government in London, England, County Hall now houses tourist attractions, business, and hotels within its grand architectural space. Situated on Southbank adjacent to the famous London Eye, County Hall uses ColorReach Powercore LED floodlights to illuminate its impressive Edwardian baroque façade to stunning effect.

ColorReach Powercore allows the building's management company to ensure that County Hall is as much of an iconic and attractive landmark by night as it is by day. The ability of ColorReach Powercore to project dynamic light shows and lighting effects also allows corporate clients who hire County Hall for private events to brand the area with company colors or other distinctive themes.

Installation involved a simple replacement of existing conventional fixtures with a total of 16 ColorReach Powercore fixtures. The fixtures were positioned at intervals between the columns to enhance the broad sweep of the crescent-

Sixteen ColorReach Powercore LED floodlights bathe the façade of London, England's County Hall in saturated, color-changing light. ColorReach Powercore ensures an iconic and attractive presence for County Hall by night, and integrates the structure with the adjacent landmark London Eye, while cutting energy consumption in half.

shaped entrance, while four fixtures were spaced alongside the façade. Each ColorReach Powercore unit consumes just 290 watts at full intensity, compared with 600 watts for the conventional fixtures that they replaced, reducing energy usage for the application by about half.

Peter Castelton, director of the building's management company, explains, "Having seen what Philips ColorReach Powercore was capable of at the initial trial we were completely convinced that this was the right solution for us . . . [I]t provides saturated colour output at significantly less cost for installation, operation and maintenance than traditional light sources — and in today's current economic climate these cost elements simply cannot be ignored."

Roadway and Area Lighting

Roadway and area lighting is designed to deliver adequate, uniform illumination for pedestrians and vehicles. Roadway lighting fixtures are typically mounted on posts or arms on a pole.

As of 2008, LED fixtures and retrofit arrays had already replaced incandescent lamps in 65% of all traffic signals and pedestrian lights in the US and Europe. As with floodlighting, LEDs offer significant energy savings in roadway lighting applications, as color filtering of incandescent lamps can block 90% or more of a fixture's light output.

The high reliability and long lifetimes of LEDs are especially crucial in roadway lighting applications, as replacing lamps is expensive and failed signals are dangerous. Beyond offering alternatives for pedestrian, street, and roadway lighting, LEDs have the potential to change the way we light our streets and roads at night. For example, LEDs can be embedded into pavements, curbs, and safety fences.

Radiant

Radiant, an LED area and roadway lighting fixture from Philips Gardco, employs a unique, prismatic optical system and high-intensity white-light LEDs for varied distribution patterns for area lighting, precise aiming, and superior light quality. Radiant can save 30% – 50% on energy as compared with many high-pressure sodium or metal halide systems, and useful life of 50,000 hours or more means less replacement and lower maintenance costs. Radiant offers state-of-the-art weatherability and construction, and a streamlined, contemporary profile suitable for a wide range of applications.

Philips Gardco Radiant roadway lighting fixtures can save up to 50% on energy costs, as compared with high-pressure sodium or metal halide fixtures.

Safety and Utility Lighting

Safety and utility lighting includes egress and exit lighting, high-bay industrial lighting, markers, dock lighting, non-explosive industrial lighting, and indicative lighting for various industries and applications.

Egress and other forms of indicative lighting can display directions and messages in a variety of circumstances. By illuminating signs and pathways, lighting helps to effectively communicate with and guide people using indoor and outdoor spaces.

Because they reliably provide bright light for many thousands of hours, LED-based safety lighting solutions can provide a level of security and reliability that conventional solutions cannot offer.

Philips Gardco Crosswalk System

The Philips Gardco Crosswalk System saves lives and prevents injuries by alerting drivers to crossing pedestrians at any hour of the day or night. The Crosswalk System features a series of in-road warning lights (IRWLs) that lie flush with the road surface and that use super-bright LEDs visible at more than 500 feet in direct sunlight and more than 1 mile at

Philips Gardco Crosswalk System uses super-bright LEDs in its in-road warning lights, which are visible at more than 500 feet in direct sunlight. Multiple control and additional warning options create an effective crosswalk solution.

night. The system controller, which offers automatic activation through a motion sensor, is housed inside an aesthetically pleasing decorative bollard. The system also offers activation through a push-button controller that can be operated by pedestrians.

The Crosswalk System is ideal for mid-block crossways, uncontrolled intersections and heavily trafficked pedestrian areas such as schools, military bases, corporate campuses and commercial areas. When activated, the system triggers both the series of IRWLs positioned in the drivers' line of site directly in front of the crosswalk, and a traditional lighted road sign for an additional level of awareness. The in-road warning lights flash at an adjustable timed interval while the pedestrians cross the street. The system can be enhanced with additional devices to alert drivers that pedestrians are present. Customized LED signs and beacons can be positioned in advance of the in-roadway components to call extra attention to the upcoming crossway and prepare drivers to yield.

The manual activation system can be powered either by a small pole-mounted AC controller, or by an energy-saving solar controller. The energy efficiency and longevity of the LEDs in this advanced system of safety lights assures cost-effective, reliable, low-maintenance operation.

Accent Lighting

Accent lighting is essentially decorative or ornamental. Sometimes the accent lighting fixture itself is designed to function as an ornament, but most often the light that the fixture produces is designed to be decorative.

Accent lighting embraces a wide range of fixture types and applications, from small spotlights for illuminating artwork, sculptures, and small architectural and landscape features, to pendants for subtle illumination in restaurants or bars or for providing decorative points of white or color-changing light, to immersible fixtures for illuminating fountains, pools, docks, and so on.

ColorBurst 6

ColorBurst® 6, from Philips Color Kinetics, is a round, color-changing LED spotlight for indoor and outdoor applications. ColorBurst 6 combines the classical look of a round, 6 in (152 mm) spotlight, providing output of over 500 lumens while consuming only 25 watts per fixture. Enclosed in a rugged, die-cast aluminum housing, ColorBurst 6 projects rich, saturated colors and color-changing effects for spotlighting, wall-washing, and accent applications. A three-screw accessory ring lets you affix spread lenses, egg-crate louvers, and other attachments for controlling and dispersing light.

ColorBurst 6 is a round, full-color LED spotlight, producing over 500 lumens while consuming only 25 watts per fixture.

iColor MR g2

iColor MR g2, from Philips Color Kinetics, demonstrates how a well-designed LED replacement lamp can offer tremendous benefits in retrofit installations. iColor MR g2 is an intelligent, full-color MR16 lamp that delivers saturated bursts of color and color-changing effects. High-intensity LEDs, two beam angles, and interchangeable clear and frosted lenses support a wide range of architectural, theatrical, and retail applications.

Standard GU5.3 base and two-pin MR16 connector make iColor MR g2 compatible with most MR16 tracks, rails, cables, and pendant fixtures. An optional adapter ring fits iColor MR g2 lamps in MR16 fixtures requiring a thin flange around the face of the lamp. iColor MR g2 lamps work with standard 2-conductor jacketed cable or hook-up wire. Power / data supplies designed for iColor MR g2 multiplex power and data onto a two-wire circuit for use with conventional MR16 fixtures and sockets.

iColor MR g2 is a full-color LED MR16 replacement lamp, bringing full-color accent and effect lighting to standard MR16 tracks, rails, cables, and pendants.

C-Splash 2

C-Splash™ 2, from Philips Color Kinetics, is a color-changing LED spotlight for use in fresh and salt water to a depth of 15 ft (4.6 m). A watertight cast brass housing and silicon bronze adjusting hardware make C-Splash 2 appropriate for water-based applications such as fountains and theme park installations, as well as for applications situated in harsh environments. Below, C-Splash 2 illuminates the Lake of Dreams at the Wynn Las Vegas resort and casino.

C-Splash 2 adapts high-output, energy-efficient LED technology to underwater accent applications, such as fountains, pools, and theme park effects.

Case Study: Accent Lighting / Hospitality Space

Costa Concordia joined the fleet of Italy's Costa Cruise line in 2006. It is the largest and longest ship in the fleet, offering all the comforts of a floating palace. Its visually striking interior features multiple applications of LED lighting technology, which enhance the meticulous detail incorporated in its stylish design.

As shown below, the ship's main atrium dazzles passengers with custom-designed chandeliers that resemble glowing, colorful sea urchins. The chandeliers are illuminated with 1,250 iColor MR g2 lamps. Each lamp

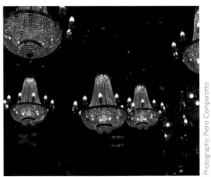

iColor MR g2 lamps add saturated color to the crystal chandeliers in the disco onboard Costa Cruise Line's Costa Concordia.

is individually controlled to generate dynamic, color-changing effects. iColor MR g2 lamps were also used to bring color to the more traditional crystal chandeliers that adorn the ship's disco, as shown above.

Over 1,200 iColor MR g2 lamps, installed in custom chandeliers in the atrium of the flagship Costa Concordia cruise ship, create dramatic, color-changing accents impossible to achieve with conventional MR16 lamps.

Case Study: Accent Lighting / Public Interior

ColorBurst 6 is designed with a fully-sealed die-cast aluminum housing for use in exterior and wet locations, but its small footprint, focused beam, and ease of installation also make it a good choice for indoor spotlighting and accent lighting.

The Richard G. Folsom Library on the Troy, New York, campus of Rensselaer Polytechnic Institute (RPI) had not been renovated since its opening in 1976. RPI recognized the need for a refurbishment that would make its main library more user-friendly and inspire and attract students, researchers, and others.

The lighting designer drafted a creative plan with an emphasis on color-changing accent lighting and spotlighting. To make the library more open and inviting, light show colors were designed to interact with the changing climate of upstate New York, displaying warm colors during the cold winter months, cool colors in the summer, greens in the spring, and orange colors in the fall.

The library uses a variety of Philips color-changing LED lighting fixtures throughout — to lend color to computer hubs and work areas, for example. ColorBurst 6 fixtures are used to accentuate an inverted world map, specially designed for the renovation, which provides a dramatic backdrop for the library's circulation desk.

A single Philips DMX controller gives library staff complete control over all of the lighting fixtures and light shows, both static and dynamic, throughout the library. In addition to providing an inviting atmosphere year-round, Philips LED lighting benefits the library through low energy consumption and minimal maintenance costs as compared with the traditional lighting sources formerly in place.

ColorBurst 6 fixtures provide inviting full-color accents to the circulation desk at Rensselaer Polytechnic Institute's Richard G. Folsom Library. A simple DMX controller gives library staff control over the light shows, allowing them to vary the colors and effects depending on the season.

Direct View Lighting

Direct-view fixtures are for looking at, not for illuminating objects or spaces. Full-color LED direct-view fixtures with fine levels of control can display complex, dynamic images and color-changing effects, as well as large-scale video. Linear and strand fixtures can create two- and three-dimensional displays on any scale, from stunning indoor set pieces to building-covering video displays visible for miles.

Because direct-view LED fixtures are different in intent from other LED fixtures, their light output is typically reported in candela and nits or lux, which measures illuminance per area. Controllers specifically designed for direct-view installations can sample and stream video to fixtures over Ethernet.

Since LED-based video can become extremely complex, many digital media and lighting design companies specialize in creating, configuring, deploying, and testing displays. These companies typically create detailed lighting plans or maps which electricians can follow to mount fixtures in the correct locations and orientations.

At Carré de Soie, a shopping and leisure center near Lyon, France, 490 flexible strands of full-color LED nodes outline a sinuous latticework entryway extending from the cinema toward the avenue. With 16,000 individually addressable nodes, the LED lighting system can display dynamic and dramatic color-changing effects.

iColor Accent Powercore

iColor Accent Powercore, from Philips Color Kinetics, is a direct-view, linear LED fixture ideally suited for creating ribbons of color and color-changing effects. Fine control resolution offers the precision to display large-scale video, graphics, and intricately designed effects in architectural, retail, and entertainment settings.

iColor Accent Powercore is available in 2 ft (610 mm), 4 ft (1.2 m), and 8 ft (2.4 m) lengths that are easily connected to create long, continuous columns of intense, dynamic color. Fixtures can be individually addressed and controlled in increments as fine as 1.2 inches (30 mm) for large-scale video and complex color-changing effects, or as large as 8 ft (2.4 m) for architectural accenting and decorating.

iColor Accent Powercore offers Ethernet-based control resolution in segments as small as 1.2 in or as large as 8 ft, for architectural accenting and building-covering video displays.

iColor Accent Powercore fixtures accept Ethernet input from a Ethernet-compatible power / data supply to support long control runs not subject to DMX data and addressing limitations.

iColor Flex LMX

iColor Flex LMX, from Philips Color Kinetics, is a flexible strand of large, high-intensity, full-color LED nodes designed for extraordinary effects and expansive installations. Each strand consists of 50 individually addressable LED nodes, featuring dynamic integration of power, communication, and control. The flexible form factor accommodates two- and three-dimensional configurations, while high light output affords superior long-distance viewing for architectural accent and perimeter lighting, large-scale signage, and building-covering video displays.

iColor Flex LMX strands offer 50 individually controllable full-color LED nodes for two- and three-dimensional effect and video installations.

iColor Flex LMX strands can be mounted directly to a surface like traditional string lights. Mounting tracks ensure straight linear runs, while single node mounts can be positioned individually to provide anchor points for installations with uneven node spacing or complex geometries.

To support different video resolutions, iColor Flex LMX strands are available with standard on-center node spacing of 4 in (102 mm) or 12 in (305 mm). Custom on-center node spacing, from 3 in (76 mm) to 24 in (610 mm), can support virtually any lighting or video design.

iColor Tile MX

iColor Tile MX is a full-color LED light panel for creating stunning light art or effect lighting in a variety of surface-mounted and recessed applications. Each 2 x 2 ft (610 x 610 mm) panel has 144 individually addressable nodes to enable fine-grained control and intricacy. iColor Tile MX is a base unit for indoor applications, ideal for wall and ceiling installations.

iColor Tile MX is a 2 x 2 ft LED light panel with 144 individually addressable nodes. It acts as a canvas for light art, effect lighting, and large-scale video displays.

Case Study: Large-Scale Video Display / Exterior Architecture

To get attention in Atlantic City and differentiate its exterior from that of its neighbors, Harrah's Resort and Casino created the world's largest outdoor video display by covering the façade of the 44-story Waterfront Tower in a linear array of over 4,500 Philips iColor Accent Powercore fixtures. Approximately 33,000 feet – almost six linear miles – of iColor Accent Powercore was used to wrap around the four sides of the Waterfront Tower.

The fixtures were installed in horizontal runs, beginning on the building's 4th floor and reaching all the way to the 44th floor. The fixtures are mounted end-to-end horizontally, spaced 9.5 feet apart vertically, and are mounted to existing mullions.

For this particular installation, the fixtures are controlled in 2.4-inch increments (or "pixels") via Video System Manager Pro, a video-based control system from Philips Color Kinetics. The project marks a milestone in the number of LED nodes that are independently controlled in a single network.

From dusk to dawn, the entire building turns into a large-scale video screen displaying scenes that range from stars, comets, and shooting stars to waving American flags and fireworks to rolling dice and scrolling card suits.

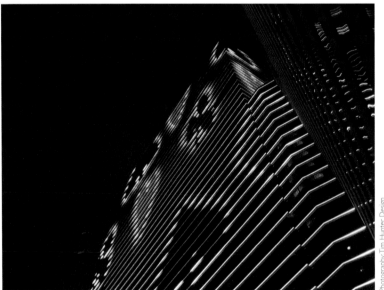

The Waterfront Tower at Harrah's Resort and Casino in Atlantic City is completely wrapped with 33,000 feet of iColor Accent Powercore fixtures, controlled to a resolution of 1.2 in to display large-scale video effects, visible for several miles.

5

Doing Business with LED Lighting

THE CURRENT FOCUS on sustainability, environmental management, and minimizing the destructive effects of human consumption continues to open up green market opportunities worldwide. The renewable energy market affords unprecedented opportunities in solar, wind, biomass, and other low-impact energy sources, while the green building market is experiencing tremendous growth in projects that use green building methods and materials selection. The largest green market by far is the energy efficiency market.

The atrium of the World Market Center in Las Vegas uses thousands of linear feet of LED cove fixtures, reducing electric load, labor, and maintenance costs for interior general illumination.

Since energy-efficient lighting is an essential component in this market's ongoing growth and success, business opportunities involving LED lighting should continue to increase sharply through 2012 and beyond.

In 2004, the total market for energy efficiency products and services was $300 billion.[34] Excluding transportation, industrial, and utilities, the buildings segment was $178 billion alone.[35] On average, 25% – 40% of all commercial energy use goes to lighting.[36] Not surprisingly, therefore, lighting retrofits and upgrades are the fastest growing segment of the energy efficiency market.[37] The US market recently received a major push from the American Recovery and Reinvestment Act of 2009 (ARRA), which provides substantial funding through 2013 for energy-efficiency retrofits and renovations for institutional and federal buildings. LED-based lighting fixtures, which offer many benefits related to energy efficiency, are well positioned to take advantage of these growing opportunities.

> ✸ In addition to LED-based retrofits and renovations, lighting upgrades for energy efficiency include daylighting controls and practices, occupancy sensors and controls, and building automation solutions such as dimming systems and load shedding.

The lighting market is about $70 billion globally. In the US, lighting represents 8% of all energy use, and about 22% of all electricity use.[38] The cost of energy for lighting is estimated to be $40 billion annually.[39] LED Lighting could reduce lighting-related energy consumption 50% by 2025. Total savings from 2000 – 2020 could reach $100 billion, and eliminate the need for more than a hundred 1000-megawatt power plants.[40]

Global LED Lighting Fixture Market Size

In 2009, the total global market for LED lighting was about $2.1 billion, 43% of which was in North America alone. This market is expected to double to $5.3 billion by 2012.[41]

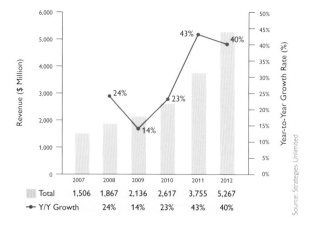

In 2009, the market for color-changing LED lighting systems was about three times as large as the market for white-light systems. This situation should change by 2012, when the market for white-light LED fixtures is expected to surpass the market for color-changing LED fixtures.[42]

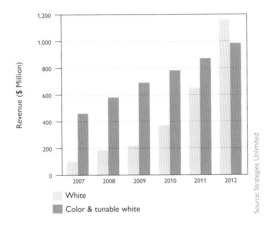

Driving the Demand: Legislation, Policies, and Incentives

So what's driving the demand for energy-efficient lighting in general, and energy-efficient white light in particular?

First, and perhaps most fundamentally, the demand for energy continues to increase, with no end in sight. A sampling of some key growth rates reveals that the global population is expected to increase 50% by mid-century to 9 billion.[43] The number of automobiles and trucks is expected to double in 30 years to 2 billion, and the number of planes operated by commercial airlines is expected to double in 20 years.[44] To fuel all this growth, 65% more energy will be required. Demand for oil alone will increase by 35%.[45]

Another important driver accelerating the demand for energy-efficient building and lighting is the momentum provided by the increasing number of successful green projects around the world and the positive publicity that these projects have generated. Measurement and analysis of many completed projects validate energy and cost savings claims, and contribute to the growing body of evidence in support of the business-case benefits of green buildings.

The 2010s promise to be a turning point in the transformation of lighting standards and practices around the world, as the global phase-out of energy inefficient lamps gains momentum. Additional initiatives, such as bans on lamps with poor color rendering and restrictions or outright bans on dozens of hazardous chemicals (notably mercury and lead) promise to further drive the move to green solutions.

Lamp replacement is only one of many possible approaches to achieving green lighting solutions, and in many ways it is the least imaginative and least effective. In new construction, renovation, and retrofit scenarios, specifying and installing complete LED-based lighting systems can deliver much higher levels of energy efficiency, flexibility, and quality of light. By combining LED lighting systems with building control systems and other automated controls, architects and lighting designers can begin to realize truly sustainable solutions, such as zero-energy buildings.

EU-27 Environmental Lighting Initiatives

Not surprisingly, the 27 member nations comprising the European Union (EU-27) are leading the way in announcing and approving energy-efficiency and other green initiatives and legislation. Since many of these initiatives were originally drafted before LED alternatives were sufficiently capable, they tend to recommend the adoption of CFLs in lamp replacement scenarios. Now that high-performance LED replacement lamps are generally available, initiatives are being revised to favor their use, since they are more energy efficient, offer longer useful lives, and contain no mercury or lead.

The following sections take a brief look at some of the major legislation, policies, and incentives helping to drive the energy-efficiency market in the EU.

Waste Electrical and Electronic Equipment Directive (WEEE)

The purpose of the WEEE directive is to reduce or prevent hazardous waste through reuse and recycling. WEEE holds manufacturers responsible for financing the environmentally sound disposal of products at end of life. As of early 2010, WEEE has been officially implemented in all EU 27 countries, as well as in Norway and Switzerland.

Under WEEE, manufacturers and importers of electronic and electrical equipment are responsible for covering the costs of collection, recovery, and treatment of electrical and electronic equipment waste. Lamp manufacturers are allowed to work together to set up collective recycling and service organizations. Consumers can deposit products free of charge at collection points administered by licensed operators.

Some lamp manufacturers, such as Philips, visibly indicate on their invoices a flat fee for financing electrical and electronic waste disposal and recycling. This fee is a significant portion of the purchase price for low-quality lamps with short rated lives, such as halophosphate fluorescent lamps.

The WEEE directive favors the use of replacement LED lamps, as they offer the longest useful life of any available lamp, and unlike CFLs they contain no mercury. Since LED lamps and fixtures are not covered by the WEEE directive, they carry no fee or surcharge for disposal.

Restriction of Hazardous Substances Directive (RoHS)

The RoHS directive, implemented in 2006 and revised in 2009, restricts the use of certain hazardous substances in electrical and electronic equipment. RoHS complements the WEEE directive by preventing end-users from coming into contact with hazardous substances, reducing risks to recycling staff, minimizing the need for special waste treatment and recycling equipment, reducing pollution, and cutting overall WEEE costs.

Hazardous substances restricted under RoHS include lead, mercury, hexavalent chromium, cadmium, polybromated biphenyls, and poybromated diphenyl ethers. RoHS offers exemptions on the restriction of certain substances for the lighting industry. Mercury is required for the efficient operation of gas discharge lamps, for example, and there are no technically acceptable alternatives to lead for glass in fluorescent tubes and high melting temperature solders.

Lighting products that use mercury and lead conform with RoHS so long as they do not exceed maximum exemption values (5 mg of mercury in CFLs, for example). As a result, the RoHS legislation effectively bans poor quality lighting products. Reputable lighting producers, such as Philips, have adopted manufacturing standards that exceed the RoHS restrictions for mercury and lead. A revised RoHS directive is expected before the end of 2010. Over time, RoHS restrictions on mercury, lead, and other hazardous substances will become stricter, and the number of exemptions will be reduced.

RoHS-like substance management policies have been or are being adopted around the world. California implemented a version of RoHS in 2007. Korea, Ukraine, and Turkey have already adopted the EU-27 RoHS regulations, and RoHS legislation is pending in China, Latin America, India, and South Africa.

Like WEEE, RoHS favors LED alternatives, as LED lighting sources contain neither mercury nor lead. As RoHS restrictions become stricter and more widespread, LED lighting alternatives to conventional lighting solutions will become increasingly attractive.

Ecodesign Directive for Energy-Related Products (ErP)

The Ecodesign Directive for Energy-related Products was extended in 2009 to all products the use of which has an impact on energy consumption, including:

- Energy-using products (EuPs) which use or generate energy (electricity, gas, fossil fuel, and so on), including consumer goods such as light bulbs.

- Energy-related products (ErPs) which do not necessarily use energy, but which have a direct or indirect impact on energy consumption, such as windows, insulation material, and automobile tires.

ErP (formerly EuP) requires manufacturers to increase the energy efficiency and reduce the negative environmental impacts of energy-related products. The ErP's implementing measures (IMs) spell out compliance standards for a range of product categories.

Commission Regulation 244/2009, the IM for non-directional household lamps, covers domestic lighting products such as incandescent, CFL, and LED lamps. IM 244 is staged to gradually phase out products that are not energy efficient by 2016. Phase 1, which went into effect in September 2009, bans the sale in the EU of most standard incandescent lamps and other lamps that are not highly energy efficient. Phase 2, expected to go into effect in September 2010, includes energy efficiency requirements for luminaires, reflector lamps, and LED lamps. When Phase 3 goes into effect in 2012, only lamps with an EU energy label classification of A, B, or C will be allowed in the EU-27 market.

Commission Regulation 245/2009, the IM for office and public building lighting, does not specifically address LEDs, but it phases out the sale of many non-LED lamps and luminaires for which highly energy-efficient, LED-based

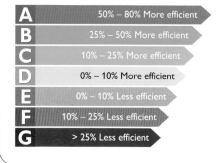

EU Energy Label Classifications for Lamps

In EU-27 member nations, light bulbs, cars, and most electrical appliances carry the EU Energy Label. Energy efficiency is rated in levels ranging from A, the most energy efficient, to G, the least energy efficient. For lamps, the levels indicate electrical consumption relative to a standard incandescent lamp that produces the same number of lumens.

A	50% – 80% More efficient
B	25% – 50% More efficient
C	10% – 25% More efficient
D	0% – 10% More efficient
E	0% – 10% Less efficient
F	10% – 25% Less efficient
G	> 25% Less efficient

alternatives are available. For example, Stage 1 phases out T8 halophosphate fluorescent lamps, and T8 and T5 lamps with CRIs below 80, in 2010. Stage 2 phases out T10 and T12 halophosphate fluorescents, and some high-pressure sodium and metal halide lamps, by 2012. Additional measures, such as increased efficiency ratings for fluorescent and HID lamp ballasts, further advantage well-design LED-based solutions. Such solutions also make business sense thanks to their energy cost savings, optimal lighting quality, lifetime reliability, and minimized use of hazardous substances.

Energy Performance of Buildings Directive (EPBD)

Originally implemented in 2006, the EPBD is part of the EU initiative to combat climate change by improving the energy performance of buildings, which account for one third of all EU energy consumption. The majority of EU-27 member states have adopted the directive under their national laws. A revised version of the EPBD is expected during 2010.

The EPBD addresses the energy consumption and efficiency of residential, office, and public buildings with a minimum of 1000 m2 of useful floor space, excluding historical buildings and industrial sites. It offers a common methodology for calculating the integrated energy performance of buildings, including all aspects of energy consumption such as heating, cooling, lighting, position and orientation of the building, heat recovery, and so on. Energy-efficient lighting, especially when integrated with building controls and other energy-efficient systems, offer strong advantages in both energy reduction and efficiency, and can help building owners and managers achieve high classification levels cost effectively.

North American Environmental Lighting Initiatives

In response to the increased demand for energy-efficient solutions, the US government has been investing heavily in green technologies. Government policies and legislation, such as ARRA, are accelerating the growth of the energy-efficiency market. As a baseline, government spending was $159 million on solar renewable energy technologies, and $25 million on solid-state lighting technologies.[46] Sustained action at state and local levels is promoting green building projects. Some cities are also beginning to require the private sector to conform to green building policies and standards.

The following sections take a brief look at some of the major legislation, policies, and incentives helping to drive the energy-efficiency market.

The Energy Policy Act (EPAct) of 2005

The Energy Policy Act (EPAct) of 2005 includes a number of resolutions related to energy efficiency, including tax incentives and loan guarantees. Among many other measures, EPAct:[47]

- Established requirements for energy savings and energy efficiency standards in federal buildings, and created a grant program to help states and local governments encourage the construction of energy-efficient public buildings.

- Created the Energy Efficient Commercial Buildings Deduction, which specifies tax deductions for reducing energy from 25% – 40% in commercial buildings, and allows building owners to deduct the cost of lighting upgrades or retrofits up to a cap of US $0.60 per square foot. The Emergency Economic Stabilization Act of 2008 extended the Energy Efficient Commercial Buildings Deduction for five years, through 2013.

- Created the Next Generation Lighting Initiative, a public-private partnership to develop advanced solid-state lighting devices. This initiative in turn led to the US Department of Energy's Solid-State Lighting Program, and strengthened the government's partnership with The Next Generation Lighting Industry Alliance (NGLIA) for accelerating advances in and the commercialization of solid-state lighting systems.

The Energy Independence and Security Act (EISA) of 2007

The Energy Independence and Security Act (EISA) of 2007 established a number of energy-efficiency standards for public buildings and lighting, including:[48]

- Around 25 percent greater efficacy for light bulbs, phased in from 2012 through 2014, effectively banning the sale of most current incandescent light bulbs and accelerating the development of new lamp technology
- Doubling the efficacy for light bulbs (or similar energy savings) by 2020
- Increased ballast efficiency standards for certain ballasts and lamps
- ENERGY STAR lighting in federal buildings, and an overall 3% reduction of energy use each year from 2008 through 2015
- Creation of a training program for energy-efficiency and renewable-energy workers
- Creation of small business energy programs that offer loans to small businesses for energy efficiency improvements

California's Title 24

In 2005 the state of California, always proactive where energy efficiency is concerned, established new energy-efficiency standards for buildings under Title 24 of the California Code of Regulations.[49]

It's important to note that the 2005 standards contained language that was insufficient for LED lighting. The standards set certain minimum efficacy requirements for LED lighting, but these requirements were set on LED sources, not on the performance of lighting fixtures containing LEDs. As we saw in Chapter 3, the operational performance of fixtures employing the same LED sources can vary widely from manufacturer to manufacturer. New standards, proposed in 2008 and in effect as of January 1, 2010, do a much better job of accounting for the performance of LED lighting fixtures.

The 2008 Title 24 standards address commercial and residential lighting separately. Among other provisions, the residential sections of the standards

Title 24 Standards for High-Efficacy LED Lighting Source Systems	
System Power Rating for LED Lighting	Minimum System Efficacy for LED Lighting
5 W or less	30 lm / W
over 5 W to 15 W	40 lm / W
over 15 W to 40 W	50 lm / W
over 40 W	60 lm / W

Source: California Energy Commission

provide guidance on the use of high-efficiency luminaires, and motion sensors and dimmers for low-efficacy luminaires. LED lighting fixtures must be explicitly certified by the California Energy Commission to be high-efficacy. If an LED "lamp" is removable from the fixture — as in the case of certain retrofit LED "lightbulbs" — it must have a GU-24 base.

Non-residential lighting changes include lower lighting power density requirements, new fixture rating methods, fewer exclusions, and new daylighting credits. LED-specific provisions require signage power supplies to be more than 80% efficient, and outdoor lighting to have multi-level switching or dimming.

ENERGY STAR for LED Lighting Program

The DoE's ENERGY STAR for LED Lighting Program, which went into effect in late 2008, sets minimum criteria for luminaire efficacy, CCT, CRI, lumen maintenance, off-state power consumption, and other important aspects of LED lighting fixture performance. Testing standards have been established for earning ENERGY STAR qualification, including IES LM-79-08, the procedures and testing standards for absolute photometry, and IES LM-80-08, the approved lumen maintenance measurement method for LED lighting fixtures.[50]

In September 2008, the first round of applications ("Category A") went into effect, including under-cabinet kitchen lighting, under-cabinet task lighting, recessed downlighting, portable desk / task lighting, and step, porch, and path lighting. In December of 2008, additional categories were added, including surface / pendant lighting, roadway lighting, outdoor decorative lighting, cove lighting, accent lighting, parking garage lighting, and bollards and troffers. The ENERGY STAR program expects to have comprehensive coverage of LED lighting fixtures for general illumination in place by 2011.

Leadership in Energy and Environmental Design (LEED)

Leadership in Energy and Environmental Design (LEED) is an internationally recognized, voluntary green building certification system developed by the US Green Building Council (USGBC). LEED uses a 100-point scale to award basic, Silver, Gold, or Platinum certification levels to commercial interiors based on evaluation of sustainability and potential environmental impacts in a number of areas.[51]

Solid-state lighting can earn points in three categories: Energy & Atmosphere, Materials & Resources, and Innovation in Design. LED sources increase LEED point totals by lowering energy consumption through their inherent efficiency, seamlessly integrating with digital building control systems, and curbing light trespass through integrated lensing for directing useful light to task areas.

Environmental Lighting Initiatives Around the World

North America and the EU are not alone in announcing and adopting energy efficiency measures that favor the use of LED and other green lighting solutions. New energy efficiency legislation, including the phase-out of incandescent lamps, is in place or underway in Australia, New Zealand, Russia, Japan, South Korea, Brazil, Argentina, Shanghai, and elsewhere. Standard TL fluorescent lamps — especially T10, T12 and halophosphate lamps — are scheduled for phase-out in the EU, Turkey, Mexico, and possibly Brazil. RoHS legislation based on the EU model is pending in the US, Latin America, South Africa, Japan, China, South Korea, and elsewhere.

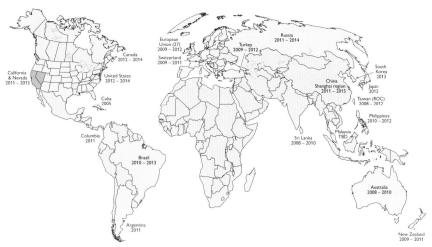

Nations around the world have banned or will soon ban incandescent lamps and low-efficiency fluorescent TL lamps. These and other green initiatives create significant market opportunities for LED lighting.

Making the Business Case

In order to successfully sell energy efficiency products, specifiers and designers must learn how to make the business case to justify upgrading to these products. At the bare minimum, they must be able to speak the language of lighting economics. The basic metrics are total cost of ownership, payback, and return on investment.

Total cost of ownership (TCO) is essentially the total cost to the owner of buying, installing, maintaining, and operating a lighting system over the total useful life of the product. LED-based lighting systems can offer a significant reduction in TCO compared to traditional technologies by offering a reduced "cost of light." This is primarily driven by the decreased cost of energy consumed by LED lighting, along with decreased maintenance expenses through useful life. Maintenance expenses include both the cost of lamps and the labor required to replace them. In addition, decreased heat generation means less heat load and lower air conditioning costs.

Payback is essentially how long it would take to recoup the incremental investment made for higher-cost, energy-efficient technology solutions. Payback is typically calculated in years, and is determined by estimating the incremental project cost and dividing by annualized savings due to decreased energy consumption and reduced maintenance. Typically, payback must be less than three years, but in today's challenging economy a payback of two years or less is more desirable.

Return on Investment (ROI) measures investment performance. Essentially, ROI is a measure of the increased profit (or decreased expense) realized by an investment, divided by the incremental cost of making that investment. In the case of an energy efficiency project, ROI would be annualized energy and maintenance savings, divided by the incremental project investment.

Cost Comparison for 10-Foot Under-Cabinet Installation			
Installation	eW Profile Powercore[1]	Halogen[2]	Xenon[3]
Number of Fixtures	6	7	6
Fixture Cost	$780	$394	$510
Number of Lamps	0	21	18
Lamp Cost	$0	$84	$126
Control Cost	$20	$20	$20
Installation ($47 / hr)[4]	$42	$263	$226
TOTAL INSTALLATION COST	$842	$761	$882
Maintenance			
Relamps per 50,000 hours	0	25	5
Lamp Cost per Change	$0	$2,100	$630
Relamp Charge[5]	$0	$419	$72
TOTAL MAINTENANCE COST	$0	$2,519	$702
Power			
Power Consumed per 10 ft (W)	60	500	330
TOTAL POWER COST[6]	$330	$2,750	$1,815
TOTAL COST OF OWNERSHIP	**$1,172**	**$6,030**	**$3,399**
Cost Savings [%]		81%	66%
Installation Premium		$81	-$39
Annual Operating Cost[7]	$19	$308	$147
Annual Savings		$288	$128
Simple Payback		0.28	-0.31
ROI		355%	-325%

1 Philips eW Profile Powercore 523-000027-01 fixture
2 WAC BA-LIV-3 fixture, 25W halogen lamps
3 Juno UPX322 fixture, 20W Xenon Lamps
4 Estimated 9 minutes per foot for LED, 0.8 hours per fixture for incandescent
5 One minute per lamp, $47 / hr labor rate
6 $0.11 / kWh, www.eia.doe.gov
7 Based on 8 hours per day usage, 365 days per year

Basic Lighting Economics Example

$$\text{Simple Payback} = \frac{\text{Incremental Investment}}{\text{Annual Savings}} = \frac{\text{LED Installation} - \text{Halogen Installation}}{\text{Annual Maintenance Savings} + \text{Annual Energy Savings}}$$

$$= \frac{\$842 - \$761}{\left[\dfrac{\$2{,}519 - \$0}{50{,}000} + \dfrac{\$2{,}750 - \$330}{50{,}000}\right] (12 \text{ hours} \times 365 \text{ days})} = 0.67 \text{ year}$$

$$\text{Simple ROI} = \frac{\text{Annual Savings}}{\text{Incremental Investment}} = \frac{1}{\text{Simple Payback}} = 150\%$$

To truly sell the value of energy-efficient lighting, you must look beyond the basic economics of lighting. Green and energy efficiency upgrades provide customers with marketing and PR opportunities in addition to reduced TCO and favorable ROI. LEED and ENERGY STAR, for example, afford strong market attraction in North America. Recent studies show a direct link between improvements in lighting quality and improvements in employee productivity, performance, and health. One Carnegie Mellon research study demonstrated a median increase in productivity of 3.2% as the result of lighting improvements. The study estimates that a 1% increase in productivity is equivalent to a 100% reduction in energy costs.[52]

Another factor to consider is employee recruitment and retention value. A Monster.com survey result found that 92% of respondents desire "sustainability" in a potential employer.[53] 80% desire an employer that is making a positive impact on the environment.[54] In addition to employee productivity and marketing opportunities studies have found an increase in building value. A CoStar market study found that on average building values increased 10 to 15 times the annual energy savings resulting from energy efficiency upgrades. For example, a $100,000 annual savings on a $300,000 energy efficiency investment would result in a three-year payback. However, it would also result in a $1.0 to $1.5 million increase in building value. That is essentially a 333% to 500% immediate return on investment.[55]

A Rapidly Changing Landscape

All signs point to a significant and sustained increase in the use of new LED lighting systems, replacement lamps, and retrofits worldwide, with the greatest growth in general illumination applications. White-light LED sources are approaching, and in some cases overtaking, conventional sources in light output and light quality, making LED lighting solutions increasingly attractive. In dozens of nations, green initiatives and energy-efficiency directives are hastening the migration from conventional lighting systems to LED lighting systems, which often deliver the lowest energy consumption and environmental impact, the longest useful life, and the lowest total cost of ownership and operation in a variety of applications.

Because LED lighting is a rapidly evolving technology, LED source and fixture suppliers must continue to leverage innovations and advances to bring the best products to market. Suppliers must also take an active role in educating lighting consumers about the specific advantages of LED lighting and how it differs from conventional lighting. Without a working understanding of LED lighting technology, consumers cannot accurately evaluate the suitability of an LED lighting system for a particular task or application, nor can they accurately compare LED lighting solutions with conventional alternatives.

LED lighting is a fundamentally new kind of lighting, using new principles, materials, and means of control. LED lighting suppliers therefore have the opportunity, if not the responsibility, to help consumers transform their understanding of lighting specification and design. LED lighting solutions can integrate and interact with people and the spaces they use in unprecedented ways. When properly evaluated and deployed, LED lighting systems have the ability to improve both the quality of the environment and the quality of life for people around the world.

Notes

1. "Facts About the Solid State Lighting Industry," August 2009. Next Generation Lighting Industry Alliance website, www.nglia.org/documents/NGLIA%20Fact%20Sheet%20August%202009.pdf, accessed November 6, 2009.

2. According to the Energy Information Administration (EIA), the U.S. residential average price for electricity in 2008 was 11.36 cents per kwh. "Short-Term Energy and Winter Fuels Outlook," October 6, 2009 release. www.eia.doe.gov/emeu/steo/pub/contents.html, accessed November 6, 2009.

3. The EIA estimates "about 526 billion kilowatt-hours (kWh) of electricity were used for lighting by the residential and commercial sectors" in the U.S. in 2007. "EIA Frequently Asked Questions – Electricity," reviewed September 21, 2009. tonto.eia.doe.gov/ask/electricity_faqs.asp#electricity_lighting, accessed November 6, 2009.

4. "The Future of Construction: Getting the U.S. There First." James M. Turner, Deputy Director of the National Institute of Standards and Technology, at the Construction Industry Institute Annual Meeting in Keystone, Colorado, August 7, 2008. Transcript at National Institute of Standards and Technology website, www.nist.gov/speeches/turner_080708.html, accessed November 6, 2009.

5. In a report prepared for the DoE in October 2008, Navigant Consulting states "if all of the previously incandescent traffic signals were converted to LED, approximately 5.2 TWh per year of electricity . . . would be saved." *Energy Savings Estimates of Light Emitting Diodes in Niche Lighting Applications*, Navigant Consulting, Inc.: Washington, D.C., 2008, p. 11.

6. Ibid., p. 70.

7. "American Recovery and Reinvestment." President Barack Obama, January, 8 2009. Transcript at Scribd website, www.scribd.com/doc/9917224/Obama-Economic-Speech-American-Recovery-and-Reinvestment-January-9-2009, accessed November 11, 2009.

8. "Cree CEO Meets President Obama to Discuss Advantages of LED Lighting." Press release at Cree Inc. website, www.cree.com/press/press_detail.asp?i=1246567403800, accessed November 11, 2009.

9. "Solid-state lighting set to boost LED growth." Rob Lineback, *LEDs Magazine*, May 2006. *LEDs Magazine* website, www.ledsmagazine.com/features/3/5/6, accessed November 11, 2009.

10. "Quality White Lighting." Philips Lumileds website, www.philipslumileds.com/technology/whitelighting.cfm, accessed November 11, 2009.

11. Rea, Mark S., ed. *The IESNA Lighting Handbook, Ninth Edition*. Illuminating Engineering Society of America: New York, NY, 2000.

12. Ibid., p. 10-13.

13. *IES Approved Method for the Electrical and Photometric Measurements of Solid-State Lighting Products*, publication IES LM-79-08. Illuminating Engineering Society: New York, NY, 2008.

14. Spec sheet for Alkco Slique T2, 3/4" Undercabinet/Display, T2 Fluorescent, SQ Series, April 2008. Philips Alkco website, www.alkco.com/upload/A10-0.pdf, accessed November 11, 2009.

15. Roscolux color filter technical data sheets, from Rosco Laboratories, Inc.
Rosco Laboratories, Inc., website, www.rosco.com/us/filters/roscolux.asp#colors, accessed November 11, 2009.

16. Rea, Mark S., ed. *The IESNA Lighting Handbook, Ninth Edition*. Illuminating Engineering Society of America: New York, NY, 2000. Pp. 10-5, 10-13, Interior-13, and Interior-16.

17. National Product Lighting Information Program. *Specifier Reports: CFL Downlights*, Vol. 3, No. 2, August 1995, and *Specifier Reports: Energy-Efficient Ceiling-Mounted Residential Luminaires*, Vol. 7, No. 2, September 1999.

18. U.S. Department of Energy. *LED Measurement Series: Color Rendering Index and LEDs*, Building Technologies Program. Publication PNNL-SA-56891, January 2008.

19. National Institute of Standards and Technology, Physics Laboratory,

Optical Technology Division. "Color Rendering of Light Sources." National Institute of Standards and Technology Physics Laboratory website, physics.nist.gov/Divisions/Div844/facilities/vision/color.html, accessed November 11, 2009.

20. American National Standards Institute. *American National Standard for electric lamps—Specifications for the Chromaticity of Solid State Lighting Products.* Publication ANSI_NEMA_ANSLG C78.377-2008, p. 8. 2008.

21. Data sheet for Golden DRAGONPlus warm white with Chip Level Conversion (CLC). OSRAM Opto Semiconductors GmbH, June 2009.

22. U.S. Department of Energy. *Energy Efficiency of White LEDs*, Building Technologies Program. Publication PNNL-SA-50462, January 2008.

23. U.S. Department of Energy. *LED Application Series: Portable Desk/Task Lighting*, Building Technologies Program. Publication PNNL-SA-54863, February 2008.

24. U.S. Department of Energy. *Thermal Management of White LEDs*, Building Technologies Program. Publication PNNL-SA-51901, April 2007.

25. Ibid.

26. Ibid.

27. "Cree® XLamp® Long-Term Lumen Maintenance," July 2009, p. 7. Cree Inc. website, www.cree.com/Products/pdf/XLampXR-E_lumen_maintenance.pdf, accessed November 11, 2009.

28. U.S. Department of Energy. *Lifetime of White LEDs*, Building Technologies Program. Publication PNNL-SA-50957, April 2007.

29. "LED Life for General Lighting: Definition of Life." *ASSIST recommends . . .* , Vol. 1, Issue 1. Alliance for Solid-State Illumination Systems and Technologies, February 2005.

30. *Life Testing of Single-Ended Compact Fluorescent Lamps*, publication IES LM-65-01. Illuminating Engineering Society: New York, NY, 2001.

31. *Measuring Lumen Maintenance of LED Light Sources*, publication IES LM-80-08. Illuminating Engineering Society: New York, NY, 2008.

32. "CALiPER Round 7 results reveal SSL progress." Brian Owen, *LEDs Magazine*, April 2009. *LEDs Magazine* website, www.ledsmagazine.com/news/6/4/16, accessed November 11, 2009.

33. Rea, Mark S., ed. *The IESNA Lighting Handbook, Ninth Edition.* Illuminating Engineering Society of America: New York, NY, 2000. P. 10-13.

34. Yudelson, Jerry and Galayda, Jaimie. *Green Goes Mainstream: How to Profit from Green Market Opportunities*. National Association of Electrical Dealers 2008 Whitepaper. NAED Education & Research Foundation, Inc., 2009.

35. Ibid.

36. Ibid.

37. Ibid.

38. *Lighting Fixtures: Industry Study with Forecasts to 2005 & 2010*. The Freedonia Group: Cleveland, OH, 2006.

39. *Energy Savings Estimates of Light Emitting Diodes in Niche Lighting Applications*. Navigant Consulting: Washington, D.C., 2008.

40. *Solid State Lighting: Brilliant Solutions for America's Energy Future*. U.S. Department of Energy, April 2009.

41. *LED Lighting Fixtures Market Review and Forecast*, section 1.4.2. Strategies Unlimited, February 2009.

42. Ibid., section 4.3.

43. *Key World Energy Statistics 2009*. International Energy Agency: Paris, France, 2009.

44. Ibid.

45. Ibid.

46. Ibid.

47. "Energy Policy Act (EPAct) of 2005." Federal Energy Regulatory Commission website, www.ferc.gov/legal/fed-sta/ene-pol-act.asp, accessed December 9, 2009.

48. "Energy Independence & Security Act." U.S Department of Energy website, www1.eere.energy.gov/femp/regulations/eisa.html, accessed December 9, 2009.

49. "California's Energy Efficiency Standards for Residential and Nonresidential Buildings." The California Energy Commission website, www.energy.ca.gov/title24/, accessed December 9, 2009.

50. "DOE's ENERGY STAR LED Lighting Program." ENERGY STAR program website, www.energystar.gov/index.cfm?c=ssl_res.pt_ssl_program, accessed December 9, 2009.

51. "LEED Version 3." U.S. Green Building Council website, www.usgbc.org/DisplayPage.aspx?CMSPageID=1970, accessed December 9, 2009.

52. Yudelson, Jerry and Galayda, Jaimie. *Green Goes Mainstream: How to Profit from Green Market Opportunities*. National Association of Electrical Dealers 2008 Whitepaper. NAED Education & Research Foundation, Inc., 2009.

53. Strandberg Consulting. *The Business Case for Sustainability*. December 2009, p. 6.

54. Mattioli, Dana. "How Going Green Draws Talent, Cuts Costs." The Wall Street Journal Digital Network website, http://online.wsj.com/article/SB119492843191791132.html, accessed March 3, 2010.

55. Burr, Andrew C. *CoStar Study Finds Energy Star, LEED Bldgs. Outperform Peers*. CoStar Realty Information, Inc. press release, March 26, 2008.

Glossary

absolute photometry
The standard method for testing the light output and light distribution of LED lighting fixtures.

additive color model
A type of RGB color model that describes how different proportions of red, green, and blue light combine to create colors. In the additive color model, combining red, green, and blue light in equal amounts produces white light.

AlGaAs
The aluminum gallium arsenide material system for manufacturing indicator-type LEDs that produce light in the yellow to orange portion of the visible light spectrum.

AlInGaP
The aluminum indium gallium phosphide material system for manufacturing red and amber high-brightness LEDs.

American National Standards Institute (ANSI)
A non-profit organization that develops voluntary consensus standards and conformity assessment systems for products, services, processes, systems, and personnel in the United States.

ANSI
See *American National Standards Institute*

ballast
　　Electronic circuitry that provides the proper electrical conditions for starting and operating fluorescent lamps and HID lamps.

binning
　　General term for the production and sorting methodologies used by LED makers to ensure that the LEDs they manufacture conform to stated specifications for forward voltage, color, and luminous flux.

black-body curve
　　A curve within a color space describing the sequence of colors emitted by a black-body radiator at different temperatures.

black body / black body radiator
　　An object that absorbs all electromagnetic radiation falling on it. Because it reflects no light, a black body appears black. As a black body is heated to incandescence, it radiates light in a sequence of colors, from red to orange to yellow to white to blue, depending on its temperature. This color sequence describes a curve within a color space, known as the black-body curve.

brightness
　　The subjective impression of the intensity of a light source. Often used incorrectly as a synonym for luminous flux, an objective measurement of the visible power of a light source.

candela
　　The standard unit of luminous intensity — power emitted by a light source in a particular direction, weighted by the spectral luminous efficiency function.

CCT
　　See *correlated color temperature*

CFL
　　See *compact fluorescent lamp*

chromaticity
　　An objective specification of the quality of a color, independent of its luminance, and as determined by its or saturation and hue.

CIE
　　See *International Commission on Illumination* (CIE = Commission internationale de l'éclairage)

CIE 1931 color space
A color space created by the International Commission on Illumination (CIE) in 1931 to define the entire gamut of colors visible to the average viewer.

color model
An abstract mathematical model describing the way colors can be represented as groups of values or color components. RGB is a color model with three color components, and CMYK is a color model with four color components.

color rendering index (CRI)
Measures the ability of a light source to reproduce the colors of various objects faithfully in reference to an ideal light source. The best possible faithfulness to the reference source has a CRI of 100.

color temperature
See *correlated color temperature*

compact fluorescent lamp (CFL)
A type of fluorescent lamp with relatively low power draw, often designed to replace an incandescent lamp.

controller
A device that controls the output of color-changing and tunable white lighting fixtures. Controllers typically have software components for configuring fixtures and designing and editing light shows, and hardware components for sending control data to fixtures.

correlated color temperature (CCT)
Describes whether white light appears warm (reddish), neutral, or cool (bluish), based on the appearance of light emitted by a black body heated to various temperatures in degrees Kelvin (K).

CRI
See *color rendering index*

DALI
See *digital addressable lighting interface*

digital addressable lighting interface (DALI)
A digital communications protocol for controlling and dimming lighting fixtures, originally developed in Europe.

delivered light
    The amount of light a lighting fixture or lighting installation delivers to a target area or task surface, measured in footcandles (fc) or lux (lx).

direct-view lighting fixtures
    Lighting fixtures intended for viewing, rather than for illumination. For example, arrays of direct-view fixtures or nodes are used in large-scale video displays.

DMX
    A digital communications protocol for controlling lighting fixtures, originally developed to control stage lighting.

efficacy
    The efficacy, or energy efficiency, of lighting fixtures is the amount of light produced (in lumens) per unit of energy consumed (in watts), or lm / W. Not to be confused with *luminous efficiency*.

efficiency
    See *luminous efficiency*. For energy efficiency, see *efficacy*.

ELV-type dimmer
    An electronic low voltage dimmer, used to dim LED lighting fixtures with electronic transformers.

energy efficiency
    See *efficacy*

Ethernet
    A digital communications protocol commonly used in computer networks. Can also be used to control Ethernet-compatible LED lighting fixtures.

eye-sensitivity curve
    See *spectral luminous efficiency function*

footcandle (fc)
    A unit of illuminance that measures the intensity of light falling on a surface area measured in square feet.

forward voltage
    Occurs when a negative charge is applied to the n-type side of a diode, allowing current to flow from the negatively-charged area to the positively-charged area.

GaAsP
>The gallium arsenide phosphide material system for manufacturing indicator-type LEDs that produce light in the red to yellow-green portion of the visible light spectrum.

GaP
>The gallium phosphide material system for manufacturing indicator-type LEDs that produce light in the green to orange portion of the visible light spectrum.

ghosting
>An effect that occurs when lighting fixtures in the OFF state faintly glow as a result of residual voltage in the circuit.

goniophotometer
>A photometric device for testing the luminous intensity distribution, efficiency, and luminous flux of luminaires.

HB-LEDs
>High-brightness LEDs. A synonym for indicator-type LEDs.

heat sink
>A feature or device that conducts or convects heat away from sensitive components, such as LEDs and electronics.

HID lamp
>A type of electrical lamp which produces light by means of an electric arc between tungsten electrodes housed inside a translucent or transparent fused quartz or fused alumina arc tube.

HP-LEDs
>High-power LEDs. A synonym for indicator-type LEDs.

IES
>See *Illuminating Engineering Society of North America*

IESNA
>See *Illuminating Engineering Society of North America*

illuminance
>The intensity of light falling on a surface area. If the area is measured in square feet, the unit of illuminance is footcandles (fc). If measured in square meters, the unit of illuminance is lux (lx).

Illuminating Engineering Society of North America (IES)
The recognized technical authority on illumination, communicating information on all aspects of good lighting practice to its members, to the lighting community, and to consumers through a variety of programs, publications, and services.

illuminator-type LEDs
High-performance, high-power LEDs capable of providing functional illumination.

inboard power integration
An approach to power management that integrates the power supply directly into a fixture's circuitry, creating an efficient power stage that consolidates line voltage conversion and LED current regulation.

indicator-type LEDs
Inexpensive, low-power LEDs suitable for use as indicator lights in panel displays and electronic devices, or instrument illumination in cars and computers.

InGaN
The indium gallium nitride material system for manufacturing green, blue, and cyan high-brightness LEDs.

integrating sphere
A photometric device for testing the total luminous flux and chromaticity (color) of a fixture's lamps.

International Commission on Illumination (CIE)
Known as the CIE from its French title, the Commission Internationale de l'Eclairage, an organization "devoted to worldwide cooperation and exchange of information on all matters relating to the science and art of light and lighting, color and vision, and image technology." (from CIE website)

infrared (IR)
Electromagnetic radiation with wavelength longer than that of visible light.

Judd-Vos modification
Adjusts the spectral luminous efficiency function so that it more accurately represent the normal sensitivity of human vision, especially to blue light.

junction
> The p-n junction in a diode, where positively charged and negatively charged materials exchange electrons, emitting photons and generating heat.

junction temperature
> The temperature in the vicinity of an LED's p-n junction. Controlling junction temperature is critical for achieving the optimal balance between lumen output and lumen maintenance.

lamp
> The light source in a luminaire or lighting fixture.

leading-edge dimmer
> A type of dimmer that regulates power to lamps by delaying the beginning of each half-cycle of AC power. Typically used with incandescent lamps.

LED
> See *light-emitting diode*

LED driver
> An electronic circuit that converts input power into a current source — a source in which current remains constant despite fluctuations in voltage. An LED driver protects LEDs from normal voltage fluctuations, overvoltages, and voltage spikes.

light-emitting diode (LED)
> A semiconductor device that emits visible light of a certain color.

light level
> See *delivered light*

light output
> See *luminous flux*

low-voltage power distribution
> A method of powering lighting fixtures that requires low-voltage power supplies or transformers and special cabling to convert line voltage into low voltage.

lumen
> The unit of measurement of luminous flux, the total energy that a light source emits across the visible wavelengths of light.

lumen maintenance
> Describes how long a light source will retain a certain percentage of its initial lumen output. For instance, L50 is the length of time a light source retains 50% or more of its initial lumen output.

lumen output
> The total lumens emitted by a lamp or luminaire.

luminance
> The amount of light emitted or reflected from a particular area, in candelas per square meter. Sometimes called *nits*.

luminous efficiency
> The percentage of total lamp lumens that a conventional lighting fixture emits, minus any blocked or wasted light. By definition, LED lighting fixtures in which the LEDs are inseparable components have a luminous efficiency of 100%.

luminous flux
> The total energy emitted by a light source across the visible wavelengths of light, measured in *lumens*.

lux (lx)
> A unit of illuminance that measures the intensity of light falling on a surface area measured in square meters.

MacAdam ellipse
> An ellipse, drawn over a color space, that defines the threshold at which a color difference becomes perceptible.

material system
> The material, such as aluminum indium gallium phosphide (AlInGaP) and indium gallium nitride (InGaN), used within an LED to produce light of a specific color.

nanometer (nm)
> The most common unit to describe the wavelength of light, equal to one billionth of a meter.

National Lighting Product Information Program (NLPIP)
> An organization, established in 1990 by the Lighting Research Center at Rensselaer Polytechnic Institute, advances the effective use of light for society and the environment by offering information on lighting products, combined from test results obtained by NLPIP and data obtained from lighting manufacturers catalogs.

n-type material
In a diode's p-n semiconductor junction, n-type material is negatively charged. Atoms in the n-type material have extra electrons.

onboard power integration
An approach to power management that integrates the power supply into a fixture's housing, eliminating the need for an external power supply.

phosphor
A coating of phosphorescent material that absorbs light from a blue or UV LED and emits most of its output in the yellow range. The proper combination of a blue or UV LED and a phosphor coating generates white light.

phosphor white
A method producing white light in a single LED by combining a short-wavelength LED, such as blue or UV, and a yellow phosphor coating.

photometry
The measurement of light in terms of its perceived brightness to the human eye. Compare *radiometry*.

photon
The basic unit of electromagnetic radiation, including visible light.

p-n junction
The location within an LED where negatively charged (n-type) material "donates" extra electrons to positively charged (p-type) material, which "accepts" them, releasing energy in the form of photons.

power factor
A measure of how effectively a device converts electric current to useful power output.

power factor correction
In an electronic device, such as an LED lighting fixture, a system of inductors, capacitors, or voltage converters to adjust the power factor of electronic devices toward the ideal power factor of 1.0.

p-type material
In a diode's p-n semiconductor junction, p-type material is positively charged. Atoms in the p-type material have electron holes — electrons missing from their outer rings.

pulse width modulation (PWM)
A method, used by most LED drivers, to regulate the amount of power to the LEDs. PWM turns LEDs on and off at high frequency, reducing total ON time to achieve a desired dimming level.

radiant flux
The total energy emitted by a light source across all wavelengths, measured in watts.

radiometry
The measurement of radiant energy (including light) in terms of absolute power. Compare *photometry*.

relative photometry
The standard method for testing the light output and light distribution of conventional lighting fixtures, where the performance of a luminaire is measured relative to the performance of its lamps.

remote phosphor
A technique that separates the phosphor from the chip in a white-light LED, improving the extraction efficiency of emitted light.

RGB color model
An additive color model in which red, green, and blue light are added together in different proportions to produce a broad range of colors, including white.

RGB white
A method of producing white light by combining the output from red, green, and blue LEDs.

SDCM
See *standard deviation of color matching*

SMDs
Surface-mount LEDs. See *illuminator-type LEDs*

spectral luminous efficiency function
A bell-shaped curve describing the sensitivity of a human eye with normal vision to the spectrum of visible light. Also known as the eye-sensitivity curve.

steradian
The standard unit of solid angle. Describes two-dimensional angular spans in three-dimensional space.

standard deviation of color matching (SDCM)
Describes the difference between two colors. A difference of one to three SDCM "steps" is virtually imperceptible, a difference of four SDCM steps is just noticeable, and a difference of more than four SDCM steps is readily visible.

subtractive color model
A color model that applies to reflective surfaces such as paints, dyes, and inks. Combining red, green, and blue in equal amounts produces black.

trailing-edge dimmer
A type of dimmer that regulates power to lamps by delaying the end of each half-cycle of AC power. Compatible with many LED fixtures.

tunable white light
White-light LED fixtures that combine channels of warm white and cool white LEDs to produce a range of color temperatures.

useful life
The length of time it takes an LED light source to reach a certain percentage of its initial lumen output. Commonly defined as lumen maintenance thresholds $L70$ (70% of initial lumen output) and $L50$ (50% of initial lumen output).

useful light
The amount of light a lighting fixture delivers in an application, minus any wasted light.

ultraviolet (UV)
Electromagnetic radiation with wavelength shorter than that of visible light.

watt (W)
The unit of radiant flux.

Acknowledgments

This book includes the contributions of many. We would like to thank the following individuals who offered guidance, drafted sections, gathered source material, and provided comments and reviews:

At Philips
Jim Anderson
Brian Bernstein
Jeff Cassis
Scott Dallaire
Mike Datta
Kevin Dowling
Chris Fournier
Lisa Giefer
Tom Hamilton
Jill Klingler
Steve Landau
Kate O'Connell
Eddy Odijk
Matt Payette
Nadya Piskun
Justin Rawlings
Tomas Sandoval
Igor Shikh
Annette Steinbusch
Mark Sterns
Rob Timmerman
Martyn Timmings

At Cree, Inc.
Mark McClear

At NJATC
Terry Coleman

At PNNL
Marc Ledbetter